SpringerBriefs in Physics

More information about this series at http://www.springer.com/series/8902

Cesareo A. Dominguez

Quantum Chromodynamics Sum Rules

 Springer

Cesareo A. Dominguez
Centre for Theoretical and Mathematical
    Physics, Department of Physics
University of Cape Town
Cape Town, South Africa

ISSN 2191-5423   ISSN 2191-5431   (electronic)
SpringerBriefs in Physics
ISBN 978-3-319-97721-8   ISBN 978-3-319-97722-5   (eBook)
https://doi.org/10.1007/978-3-319-97722-5

Library of Congress Control Number: 2018949890

This Springer imprint is published by the registered company Springer Nature Switzerland AG
The registered company address is: Gewerbestrasse 11, 6330 Cham, Switzerland

*To Pavel Baikov, Konstantin Chetyrkin,
and Johann Kühn*

*for obtaining higher order QCD results
allowing for precision determinations from
QCD Sum Rules*

Preface

This book is intended for readers with a good knowledge of quantum field theory, in general, and quantum chromodynamics (QCD), in particular. It is addressed to readers planning to start research in QCD in the framework of sum rules. Currently, there are two major approaches to obtain information in QCD, i.e. lattice QCD (LQCD) and QCD sum rules (QCDSR). The latter is the subject matter of this book. It deals with the current state-of-the-art formulation of QCDSR in the complex squared energy plane, called finite energy sum rules (FESR). This allows for a relation between QCD and hadronic physics following from Cauchy's residue theorem in that plane. As a result, current FESR determinations of a plethora of QCD and hadronic parameters rival in precision with those from LQCD. This healthy competition is extremely beneficial for our understanding of the strong interactions at the most elementary level.

This book is not a review of past work on QCDSR. The pioneering formulation of QCDSR in the framework of integral transforms, e.g. Laplace and Hilbert, while having played a fundamental role in the development of the subject, is currently no match for the precision achieved from FESR. In addition, and most importantly, the FESR parameter related to quark-gluon deconfinement (at finite temperature) has recently been shown to be related to the Polyakov loop of LQCD. This brings these two approaches into a beneficial partnership.

The topics discussed in this book concern mostly QCD at zero temperature. A last chapter on finite temperature QCDSR has been kept short, as there is a recent comprehensive review on this subject. The extension of QCDSR to include hadronic/QCD matter in the presence of very strong magnetic fields is not covered here. This new research direction is currently in a state of flux, so the reader is advised to consult the literature.

Cape Town, South Africa   Cesareo A. Dominguez

Acknowledgements

The author wishes to thank his QCD sum rule collaborators: Alejandro Ayala, Jose Bordes, Pietro Colangelo, Marcelo Loewe, Giuseppe Nardulli[†], Nasrallah Nasrallah, Nello Paver, Jose Peñarrocha, Eduardo de Rafael, J. Cristobal Rojas, Karl Schilcher, Joan Sola, Hubert Spiesberger, and Cristian Villavicencio.

Special thanks are due to Marcelo Loewe for reading the manuscript, and to Alexes Mes and Jed Stephens for providing Eq. (3.31).

This work was supported in part by the University of Cape Town, (South Africa), and by the Alexander von Humboldt Foundation (Germany).

Contents

Chapter 1
Introduction

The theory of Quantum Chromodynamics (QCD), i.e. its Lagrangian together with its main features, was first proposed by Harald Fritzsch and Murray Gell-Mann in 1972 [1], almost a year before the discovery of asymptotic freedom [2] in QCD. The Lagrangian is

$$\mathcal{L}_{QCD} = i\,\bar{\psi}_a(x)\,\gamma_\mu\,\partial^\mu\,\psi_a(x) \; - \; m_0\,\bar{\psi}_a(x)\,\psi_a(x) - \frac{1}{4}F^i_{\mu\nu}(x)F^{\mu\nu}_i(x)$$

$$- \; g\,G_{i\mu}(x)\,\bar{\psi}_a(x)\,\gamma^\mu\,\lambda^i_{ab}\,\psi_b(x)\,, \tag{1.1}$$

where $a = 1, 2, N_c = 3$ is the SU(3)-colour index, $i = 1, 2, ...8 \equiv N_c^2 - 1$, $\psi(x)$ are the quark fields, $G_{i\mu}$ is the gluon field, λ^i the SU(3) Gell-Mann matrices, and $F^i_{\mu\nu}$ the gluon tensor

$$F^i_{\mu\nu}(x) \equiv \partial_\mu\,G^i_\nu(x) - \partial_\nu\,G^i_\mu(x) - gf_{ijk}\,G^j_\mu(x)\,G^k_\nu(x)\,, \tag{1.2}$$

where the f_{ijk} are proportional to the commutator of the SU(3) Gell-Mann matrices, $[\lambda_i, \lambda_j] = 2\,i f_{ijk}\,\lambda_k$.

In addition to the exact $SU(3)_c$ colour symmetry, with $N_c^2 - 1$ massless gluons, there is a non-trivial underlying symmetry hierarchy, incorporating several concepts from Current Algebra, a pre-QCD attempt to understand strong interactions [3].

To begin with, a label must be introduced to differentiate the quark fields according to flavour, i.e. $\psi^A(x)$, where $A =$ up, down,...etc., and the colour label is to be understood. Considering the light-quark sector, up-, down-, and strange-quarks, there are two different types of symmetries in QCD, a flavour symmetry and a chiral symmetry, depending on whether it manifests itself in the states (irreducible representations of the symmetry group) or not. The former is a classification symmetry, the latter a dynamical one with rich consequences. Starting with the flavour symmetry in the limit $m_u = m_d = m_s = 0$, i.e. $SU(3)_f$, it is realized in the

© The Author(s), under exclusive licence to Springer Nature Switzerland AG 2018
C. A. Dominguez, *Quantum Chromodynamics Sum Rules*,
SpringerBriefs in Physics, https://doi.org/10.1007/978-3-319-97722-5_1

Wigner-Weyl mode. This means that physical states (hadrons) are classified according to the irreducible representations of the group, with the vacuum sharing the Lagrangian symmetry. The first symmetry-breaking step is to make $m_s \neq 0$, with $m_u = m_d = 0$, thus breaking $SU(3)_f$, but preserving $SU(2)_f$. The next step involves $m_u = m_d \neq 0$, which still respects $SU(2)_f$, as the divergence of the vector current still vanishes, i.e. $\partial^\mu V_\mu(x) \propto (m_d - m_u) = 0$. In the final step $m_u \neq m_d \neq 0$ breaks $SU(2)_f$ down to U(1). This symmetry (eightfold-way [4]) allows for the classification of hadronic states into multiples, leads to mass formulas, and led to the prediction of the Ω^- baryon of mass 1686 MeV, discovered in 1964 [5].

Turning to chiral symmetries, an axial-vector current must be considered together with the vector current, i.e. $A_\mu(x) = \bar{\psi}(x)\gamma_5\gamma_\mu\psi(x)$. This current cannot generate a group by itself, as the commutator of two axial-vector currents transforms as a vector current. Hence, the chiral symmetry group becomes $SU(3) \times SU(3)$, with $\partial^\mu A_\mu(x) \propto (m_s + m_{ud})$, and $m_{ud} = m_u + m_d$. The question is how is this chiral symmetry realized. A Wigner-Weyl realization would imply, among other things, the existence of quasi-degenerate parity doublets not seen in the spectrum. The alternative, a Nambu-Goldstone (NG) realization of chiral symmetry, implies a massless NG-boson, plus a plethora of dynamical relations among hadronic quantities. The NG-boson is identified with the electrically charged kaon, after $SU(3) \times SU(3)$ breaking, and with the charged pion after the breaking of $SU(2) \times SU(2)$.

One of the major consequences of a NG realization of chiral symmetry is the Goldberger-Treiman relation (GTR) [6, 7]

$$\sqrt{2}f_\pi \, g_{\pi NN} = (M_p + M_n) \, g_A \,, \tag{1.3}$$

where $f_\pi = 92.28 \pm 0.07$ MeV [8] is the pion decay constant

$$\langle 0|A_\mu(0)|\pi(p)\rangle = i\,\sqrt{2}f_\pi \, p_\mu \,, \tag{1.4}$$

$g_{\pi NN} \simeq 13$ is the strong coupling, and $g_A \simeq 1.3$ the weak beta decay constant. The GTR is currently satisfied at the 1% level. Alternatives to this realization of chiral symmetry and pattern of its breaking, proposed in the past, have not survived stringent tests from $\pi - -\pi$ scattering [9].

Another crucial consequence of chiral symmetry is the Gell-Mann–Oakes–Renner relation (GMOR), first obtained in the framework of Current Algebra [10], prior to QCD, and now understood as a consequence of the NG realization of chiral symmetry. An outline of its derivation is important at this stage, as it requires the introduction of the concept of current correlators, the fundamental objects in the QCD sum rule (QCDSR) programme.

The starting point is the concept of a current correlator in momentum space, $\psi_5(q^2)$, defined in QCD as the Fourier transform

$$\psi_5(s \equiv -q^2) = i\int d^4x \, e^{iqx} < 0|\,T(j_5(x)\,j_5(0))\,|0 > \,, \tag{1.5}$$

where $< 0|$ is the physical vacuum and the current density $j_5(x)$ is chosen as

$$j_5(x) = (m_d + m_u) : \overline{d}(x) \, i \, \gamma_5 \, u(x) : , \tag{1.6}$$

with $m_{u,d}$ the light quark masses, and a convenient change in notation for the quark fields. Notice that the current densities, $j_5(x)$, are actually the QCD axial-vector current divergences

$$\partial^\mu A_\mu(x)|_{QCD} = (m_u + m_d) : \overline{d}(x) \, i \, \gamma_5 \, u(x) : . \tag{1.7}$$

Current correlators have also an alternative representation in terms of hadronic degrees of freedom, obtained by using hadronic, instead of QCD fields. For instance, in this particular case of $\psi_5(s)$, the lowest hadronic state is the pion, so that differentiating Eq. (1.4) gives

$$\partial^\mu A_\mu(x)|_{HAD} = \sqrt{2} f_\pi \, M_\pi^2 \, \phi_\pi(x) , \tag{1.8}$$

where $\phi_\pi(x)$ is the pion field, and hadronic excited states of the pion can be neglected at this stage.

The pseudoscalar current correlator, Eq. (1.5), satisfies the following low energy theorem, first obtained in the $SU(2) \times SU(2)$ Current Algebra framework [10], valid in QCD at leading order in the quark masses

$$\psi_5(0) = -(m_u + m_d) \, \langle 0 | \overline{u} \, u + \overline{d} \, d | 0 \rangle + \mathcal{O}(m_{u,d}^2) , \tag{1.9}$$

where the quark masses and fields depend on the renormalization scale, but their product, as above, does not!

An important consequence of the NG realization of $SU(2) \times SU(2)$ is that in the symmetry limit the pion mass squared vanishes as the (light) quark mass

$$M_\pi^2 = B \, m_q , \tag{1.10}$$

and the pion decay constant squared is proportional to the quark-condensate, which only vanishes at finite temperature

$$f_\pi^2 = -\frac{1}{B} \langle \overline{q} \, q \rangle . \tag{1.11}$$

These two relations have important consequences in phenomenology. In particular, in the extension of the QCDSR programme to finite temperature, as well as strong magnetic fields.

So far one has two different representations of the same object, the current correlator within QCD, and in the hadronic sector. The essential question is how to relate them. The answer, the first pillar of QCDSR, is provided by considering the complex squared energy s-plane, and invoking Cauchy theorem, as first proposed in

Fig. 1.1 Integration contour in the complex squared energy s-plane. The discontinuity across the real axis brings in the hadronic spectral function, while integration around the circle involves the QCD correlator. The radius of the circle is s_0, the onset of QCD

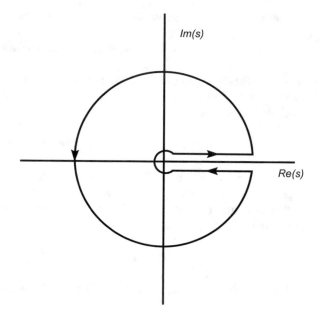

[11] (see also [12]). In QCD there are no singularities in this plane, except on the real positive s-axis, where poles correspond to stable hadrons, and on the second Riemann sheet corresponding to resonances of certain width Γ, with $\Gamma \propto 1/\tau$, and τ their lifetime. Next, an integration contour in this plane, a circle of radius $|s_0|$, is considered as in Fig. 1.1. While QCD is not valid on the real axis, it is expected to hold everywhere else on the circle, provided the radius $|s_0|$ is large enough. Cauchy theorem then relates the Physics on the real axis to that on the circle. This approach, named quark-hadron duality, leads to QCD Finite Energy Sum Rules (FESR). Since QCD is not valid on the real axis, it has been customary to multiply spectral functions by kernels vanishing on the real axis (pinched kernels) [13, 14], in order to quench this contribution. This procedure is usually quite satisfactory.

In detail, if $\Pi(s)$ is some meromorphic correlation function in the complex s-plane, and $P(s)$ a meromorphic integration kernel, Cauchy theorem states

$$\frac{1}{2\pi i} \oint_{C(|s_0|)} \Pi(s) \, P(s) \, ds = \sum_i R_i , \tag{1.12}$$

where R_i are the residues at the poles of $\Pi(s)P(s)$. Splitting the integration range into the circle and the real axis gives the FESR

$$\frac{1}{2\pi i} \oint_{C(|s_0|)} ds \, \Pi(s)_{\text{QCD}} \, P(s) + \int_{s_{th}}^{s_0} ds \, \frac{1}{\pi} \, \text{Im} \, \Pi(s)_{\text{HAD}} \, P(s) = \sum_i R_i. \tag{1.13}$$

The low energy theorem, Eq. (1.9), can be used together with this FESR, for $\Pi(s) \equiv \psi_5(s)/s$, and the hadronic spectral function, Eq. (1.8), to obtain the Gell-Mann-Oakes-Renner (GMOR) relation [10]

$$\psi_5(0) \equiv -(m_u + m_d) \langle 0|\bar{u}u + \bar{d}d|0 \rangle = 2f_\pi^2 M_\pi^2 (1 - \delta_\pi), \qquad (1.14)$$

valid to leading order in the quark masses and in the hadronic spectral function, and δ_π encapsulates higher order hadronic corrections. These are at the level of 7% in $SU(2) \times SU(2)$ [15], and 50% in $SU(3) \times SU(3)$ [16].

The GMOR relation, Eq. (1.14), is unique in the sense that an expression for the quark condensate is obtained from first principles. Other vacuum condensates, entering the Operator Product Expansion of current correlators at short distances, do not share this feature. They must be determined from the FESR themselves, or from Lattice QCD (LQCD).

In the framework of FESR the crucial issues are the value of the radius, $|s_0|$, its impact on predicted quantities, and the so-called duality violations due to the fact that QCD is not valid on the positive real axis. Regarding $|s_0|$, its uncertainty impacts on results at a very reasonable level, and combines favourably with other uncertainties in QCD and hadronic parameters. Concerning duality violations, these are unknown by definition [17, 18], thus requiring specific models. However, it is expected that the use of pinched kernels [13, 14], together with large enough radii in the complex squared energy plane, quenches substantially their importance.

Concerning the contour integral around the circle of radius s_0 in Eq. (1.13), there are two different procedures to perform this integration [19]. Both make use of the Renormalization Group Improvement (RGI) of the perturbative expansion of the QCD correlator. In the first one, called Fixed Order Perturbation Theory [20], the contour integral is performed followed by the RGI. The second procedure, Contour Improved Perturbation Theory (CIPT) [21], requires RGI to be performed first, followed by the contour integration. It is to be noticed that in FOPT the strong coupling is frozen at $s = |s_0|$, while in CIPT $\alpha_s(s)$ is to be integrated around the circle. The latter is usually performed by solving numerically the renormalization group equation for $\alpha_s(s)$ at each point on the circle.

Historically, QCDSR were first developed in the framework of the Laplace transform [22], involving an ad-hoc parameter needed for dimensional reasons in the Laplace exponential kernel. This free parameter comes in addition to the threshold for perturbative QCD, s_0, needed as a cut-off in the integrals. With s_0 having a clear physical interpretation, it is unfortunate for it to be exponentially suppressed in this approach. A serious drawback of this technique is that at and beyond next-to-next to leading order in perturbative QCD the Laplace transform leads to a Volterra-type function $\mu(x, \beta)$ [23], a fact discovered only when higher orders in perturbation theory were first determined [24]. This turns higher order perturbative corrections into a cumbersome exercise (see e.g. [25]).

In any case, a large number of applications of these sum rules were developed over the years, as reviewed in [26], leading to results affected by uncontrollable systematic uncertainties. For a critical detailed discussion of these sum rules see [27]. As high precision was eventually required in some instances, e.g. quark mass determinations, this method has been superseded by the FESR technique.

References

1. H. Fritzsch, M. Gell-Mann, in *Proceedings of the XVI International Conference on High Energy Physics, Chicago*, vol. 2, eds by J.D. Jackson, A. Roberts (1972), pp. 135. arXiv:hep-ph/0208010
2. H.D. Politzer, Phys. Rev. Lett. **30**, 1346 (1973) (Phys. Rep. **14** C, 274 (1974); D. Gross, F. Wilczek, Phys. Rev. Lett. **30**, 1343 (1973))
3. S.B. Treiman, R. Jackiw, B. Zumino, E. Witten, *Current Algebra and Anomalies*, World Scientific, Singapore (1985). (See also: J. Bernstein, *Elementary Particles and their Currents* (W. H. Freeman & Co., San Francisco, 1968)
4. The "Eightfold Way" was never published in a journal. For a historical account see e.g. G. Zweig, in *Proceedings of the Conference in Honour of Murray Gell-Mann's 80th Birthday*, eds by H. Fritzsch, K.K. Phua (World Scientific, Singapore, 2011)
5. V.E. Barnes et al., Phys. Rev. Lett. **12**, 204 (1964)
6. H. Pagels, Phys. Rep. **16**, 219 (1975)
7. C.A. Dominguez, Phys. Rev. D **7**, 1252 (1973). (ibid. D **16**, 2320 (1977); Riv. Nuovo Cim. **8N6**, 1 (1985))
8. K.G. Patrignani et al., Particle Data Group. Chin. Phys. C **40**, 100001 (2016)
9. J. Stern, "*Light quark masses and condensates in QCD*", in Mainz 1997, Chiral Dynamics: Theory and Experiment, eds. A. Bernstein, D. Drechsler and Th. Walcher [hep-ph/9712438]; J. Stern, in Chiral Dynamics: Theory and experiment III, eds. by A.M. Bernstein, J.L. Goity, and U. Meissner. World Scientific, Singapore, 2001
10. M. Gell-Mann, R.J. Oakes, B. Renner, Phys. Rev. **175**, 2195 (1968)
11. R. Shankar, Phys. Rev. D **15**, 755 (1977)
12. A. Bramon, E. Etim, M. Greco, Phys. Lett. B **41**, 609 (1972). (M. Greco, Nucl. Phys. B **63**, 398 (1973); E. Etim, M. Greco, Lett. Nuovo Cim. **12**, 91 (1975))
13. K. Maltman, Phys. Lett. B **440**, 367 (1998)
14. C.A. Dominguez, K. Schilcher, Phys. Lett. B **581**, 193 (2004)
15. J. Bordes, C.A. Dominguez, P. Moodley, J. Penarrocha, K. Schilcher, J. High Energy Phys. **1005**, 064 (2010)
16. J. Bordes, C.A. Dominguez, P. Moodley, J. Penarrocha, K. Schilcher, J. High Energy Phys. **1210**, 102 (2012)
17. M. Gonzalez-Alonso, A. Pich, J. Prades, Phys. Rev. D **81**, 074007 (2010)
18. A. Pich, A. Rodriguez-Sanchez, Mod. Phys. Lett. A **31**, 1630032 (2016)
19. M. Jamin, J. High Energy Phys. **09**, 058 (2005)
20. A.A. Pivovarov, Z. Phys. C **53**, 461 (1992)
21. F. LeDiberder, A. Pich, Phys. Lett. B **286**, 147 (1992)
22. M.A. Shifman, A.I. Vainshtein, V.I. Zakharov, Nucl. Phys. B **147**, 385 (1979). (ibid., B **147**, 448 (1979))
23. A. Erdelyi (ed.), *Higher Trascendental Functions*, vol. 3 (McGraw-Hill Book Company, Inc. New York, Toronto, London, 1955)
24. E. de Rafael, in Marseille 1981. In: J.W. Dash (ed.), Theoretical Aspects of Quantum Chromodynamics, Centre de Physique Theorique, Marseille, France, CPT-81/P.1345, 259 (1981)
25. K.G. Chetyrkin, C.A. Dominguez, D. Pirjol, K. Schilcher, Phys. Rev. D **51**, 5090 (1995). (K.G. Chetyrkin, D. Pirjol, K. Schilcher, Phys. Lett. B **404**, 337 (1997); C.A. Dominguez, L. Pirovano, K. Schilcher. Phys. Lett. B **425**, 193 (1998))

26. P. Colangelo, A. Khodjamirian, in *At the Frontier of Particle Physics*, vol. 3, ed. by M. Shifman (World Scientific, Singapore 2001), pp. 1495–1576
27. C. A. Dominguez, Analytical determination of QCD quark masses, in *Fifty Years of Quarks*, eds. by H. Fritzsch, M. Gell-Mann (World Scientific Publishing Co., Singapore 2015), pp. 287–313

Chapter 2
Operator Product Expansion in QCD

The second of the two pillars of the QCDSR method is the Operator Product Expansion (OPE) of current correlators at short distances, beyond perturbation theory. This allows for a current correlator to be written in terms of QCD degrees of freedom, starting with its perturbative expansion, followed by non-perturbative terms. The latter involve vacuum expectation values of quark and gluon fields, reflecting colour confinement. An example is that of the quark condensate, discussed in the previous section. Given some QCD correlation function

$$\Pi(q^2)|_{QCD} = i \int d^4x \, e^{iqx} < 0| \, T(J^\dagger(x) \, J(0)) \, |0 > , \qquad (2.1)$$

where $J(x)$ is some local current built from the QCD fields, the OPE is given by

$$\Pi(q^2)|_{QCD} = C_0(q^2, \mu^2) \, \hat{I} + \sum_{N=0} C_{2N+2}(q^2, \mu^2) \, \langle 0| \hat{O}_{2N+2}(\mu^2)|0 \rangle , \qquad (2.2)$$

where μ^2 is some renormalization scale and the Wilson coefficients, C_{2N+2}, depend on the Lorentz indices and quantum numbers of $J(x)$ and of the local gauge invariant operators $\hat{O}_N$ built from the quark and gluon fields. These operators are ordered by increasing dimensionality, and the Wilson coefficients, calculable in PQCD, fall off by corresponding powers of $-q^2$. Hence, the OPE implies a factorization of short distance effects, encapsulated in the Wilson coefficients, and long distance dynamics from the vacuum condensates. The term $C_0 \, \hat{I}$ stands for the purely perturbative contribution, currently known for some correlators up to order $\mathcal{O}(\alpha_s^6)$, with $\alpha_s \equiv \alpha_s(q^2)$ the strong coupling. Figure 2.1 illustrates this contribution to next-to-leading order. In QCD there are no gauge-invariant operators of dimension $d = 2$, other than light-quark mass terms, $\mathcal{O}(m_q^2)$, usually negligible.

© The Author(s), under exclusive licence to Springer Nature Switzerland AG 2018
C. A. Dominguez, *Quantum Chromodynamics Sum Rules*,
SpringerBriefs in Physics, https://doi.org/10.1007/978-3-319-97722-5_2

Fig. 2.1 Schematic representation of the purely perturbative QCD term, $C_0 \hat{I}$ in Eq. (2.2), with a quark loop and with one-gluon exchange

Fig. 2.2 Schematic representation of the QCD quark-condensate, dimension $d = 3$ term in the OPE, $\langle 0|\bar{q}(0)\, q(0)|0\rangle$, originating from on-shell light quarks (zero-momentum) interacting with the QCD vacuum. Large momentum flows through the bottom propagator

Next, there is no analytic, first-principles approach to determine the vacuum condensates in the OPE, Eq. (2.2), except for the quark condensate. This is known from the GMOR relation discussed in Chap. 1. In order to see the emergence of the vacuum condensates let us consider the quark propagator

$$ S_F(p) = \frac{i}{\not{p} - m} \implies \frac{i}{\not{p} - m + \Sigma(p^2)} , \tag{2.3} $$

where $\Sigma(p^2)$ encapsulates a correction due to confinement, which is not calculable analytically from first principles. Nevertheless, this correction is expected to peak at the quark mass-shell. For light quarks this would be $p \simeq 0$, leading to these quarks condensing into a condensate, $\langle 0|\bar{q}(0)\, q(0)|0\rangle$, as shown in Fig. 2.2

In the case of the gluon propagator

$$ D_F(k) = \frac{i}{k^2} \implies \frac{i}{k^2 + \Lambda(k^2)} , \tag{2.4} $$

the correction term, $\Lambda(k^2)$, peaks at $k \simeq 0$, as the gluon is massless. Confinement information is then parametrized by $\langle 0|\alpha_s\, \vec{G}^{\mu\nu} \cdot \vec{G}_{\mu\nu}|0\rangle$, the gluon condensate, as shown in Fig. 2.3.

Fig. 2.3 Schematic
representation of the QCD
gluon-condensate, $d = 4$
term, in the OPE. Large
momentum flows through
the quark propagators

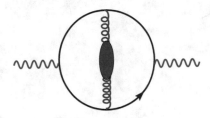

Higher dimensional condensates can be constructed from the quark and gluon
fields in the QCD Lagrangian, but their values are affected by very large uncertainties.
Due to this limitation one normally considers FESR up to dimension $d = 4$, thus
avoiding these higher-dimensional condensates.

In the heavy-quark sector, i.e. charm and bottom, the heavy-quark condensate,
while present, is unrelated to symmetry. Instead, it is related to the gluon condensate
through [1]

$$\langle \bar{Q}Q \rangle = -\frac{1}{12\,m_Q} \left\langle \frac{\alpha_s}{\pi}\, G^2 \right\rangle + \mathcal{O}\left(\frac{1}{m_Q^2}\right). \tag{2.5}$$

Reference

1. M. A. Shifman, A. I. Vainshtein, V. I. Zakharov, Nucl. Phys. B **147**, 385 (1979). (ibid., B **147**, 448 (1979))

Chapter 3
Renormalization Group Equation

The momentum dependence of the QCD coupling and of the quark masses is needed in most applications of QCDSR. The QCD strong coupling, $\alpha_s \equiv g^2/4\pi$, with g in Eq.(1.1), satisfies the renormalization group equation

$$s \frac{d}{ds} a_s(s) = \beta[a_s(s)], \tag{3.1}$$

where $a_s \equiv \alpha_s/\pi$, and the beta function is

$$\beta[a_s(s)] = -\sum_{N=0} b_N \, a_s(s)^{N+2}. \tag{3.2}$$

The expansion coefficients, b_N, are [1–3]

$$b_0 = \frac{1}{4}\left(11 - \frac{2}{3}n_f\right). \tag{3.3}$$

$$b_1 = \frac{1}{16}\left(102 - \frac{38}{3}n_f\right). \tag{3.4}$$

$$b_2 = \frac{1}{64}\left(\frac{2857}{2} - \frac{5033}{18}n_f + \frac{325}{54}n_f^2\right). \tag{3.5}$$

$$b_3 = \frac{1}{4^4}\left(\frac{149753}{6} + 3564\,\zeta(3) - \left(\frac{1078361}{162} + \frac{6508}{27}\zeta(3)\right)n_f\right.$$
$$\left.\left(\frac{50065}{162} + \frac{6472}{81}\zeta(3)\right)n_f^2 + \frac{1093}{729}n_f^3\right). \tag{3.6}$$

$$b_4 = \frac{1}{4^5} \left\{ \frac{8157455}{16} + \frac{621885}{2} \zeta(3) - \frac{88209}{2} \zeta(4) - 288090 \, \zeta(5) \right.$$

$$+ n_f \left(-\frac{336460813}{1944} - \frac{4811164}{81} \zeta(3) + \frac{33935}{6} \zeta(4) + \frac{1358995}{27} \zeta(5) \right)$$

$$+ n_f^2 \left(\frac{25960913}{1944} + \frac{698531}{81} \zeta(3) - \frac{10526}{9} \zeta(4) - \frac{381760}{81} \zeta(5) \right)$$

$$+ n_f^3 \left(-\frac{630559}{5832} - \frac{48722}{243} \zeta(3) + \frac{1618}{27} \zeta(4) + \frac{460}{9} \zeta(5) \right)$$

$$+ n_f^4 \left(\frac{1205}{2916} - \frac{152}{81} \zeta(3) \right) \right\}. \tag{3.7}$$

The values of the Riemann theta-function, $\zeta(n)$, are $\zeta(2) = \frac{\pi^2}{6}$, $\zeta(3) = 1.2020569032$, $\zeta(4) = \frac{\pi^4}{90}$, $\zeta(5) = 1.0369277551$, $\zeta(6) = \frac{\pi^6}{945}$, $\zeta(7) = 1.0083492774$, $\zeta(8) = \frac{\pi^8}{9450}$.

It is important to notice that there is an alternative definition of the coefficients b_N in the literature, i.e. $\bar{b}_N = -\frac{1}{2} b_{N+1}$. Another important observation is that the coefficients b_n are usually referred to as β_n in the literature. This could lead to confusion as the symbol β would appear on both sides of Eq. (3.2).

A very useful way to solve Eq. (3.1) is by Taylor-series developing $a_s(s)$ about some reference scale $s = s^*$, at which $a_s(s^*)$ is well known. For instance, a_s is known with high precision at the tau-lepton scale [4]

$$\alpha_s(M_\tau^2) = 0.328 \pm 0.013. \tag{3.8}$$

The expansion of $a_s(s)$ in terms of $a_s(s^*)$ reads [5]

$$a_s(s) = a_s(s^*) + [a_s(s^*)]^2 \, (-b_0 \, \eta) + [a_s(s^*)]^3 (-b_1 \, \eta + b_0^2 \, \eta^2)$$

$$+ [a_s(s^*)]^4 \left(-b_2 \, \eta + \frac{5}{2} b_0 b_1 \, \eta^2 - b_0^3 \, \eta^3 \right)$$

$$+ [a_s(s^*)]^5 \left(-b_3 \eta + \frac{3}{2} b_1^2 \eta^2 + 3 b_0 b_2 \eta^2 - \frac{13}{3} b_0^2 b_1 \eta^3 + b_0^4 \eta^4 \right)$$

$$+ [a_s(s^*)]^6 \left(-b_4 \eta + \frac{7}{2} b_0 b_1 \eta^2 + \frac{7}{2} b_0 b_3 \eta^2 - \frac{35}{6} b_0 b_1^2 \eta^3 - 6 b_0^2 b_2 \eta^3 \right.$$

$$+ \frac{77}{12} b_0^3 b_1 \eta^4 - b_0^5 \eta^5 \Big), \tag{3.9}$$

where $\eta \equiv \ln(s/s^*)$. This procedure avoids the use of the scale Λ_{QCD} in the expression

$$\alpha_s(s) \propto \frac{1}{\ln[s/\Lambda_{QCD}^2]} \cdots , \qquad (3.10)$$

as Λ_{QCD} is affected by a large uncertainty [6]. Equation (3.9) can be used e.g. to obtain a_s at the light-quark sector scale using its very precise value at the Z-boson scale, $a_s(s^* = M_Z^2)$.

Next, the renormalization group equation for the quark masses, $m_q(t)$, with $t = \ln(-q^2/\mu^2)$, is given by

$$\frac{1}{m_q(t)} \frac{d}{dt} m_q(t) = \gamma[a_s(t)], = -a_s(t) \sum_{N=0} \gamma_N \, a_s(t)^N , \qquad (3.11)$$

where the coefficients γ_i are [1–3, 7, 8]

$$\gamma_0 = 1 \qquad (3.12)$$

$$\gamma_1 = \frac{1}{16} \left(\frac{202}{3} - \frac{20}{9} n_f \right) \qquad (3.13)$$

$$\gamma_2 = \frac{1}{64} \left(1249 - \left(\frac{2216}{27} + \frac{160}{3} \zeta_3 \right) n_f - \frac{140}{81} n_f^2 \right) \qquad (3.14)$$

$$\gamma_3 = \frac{1}{256} \left\{ \frac{4603055}{162} + \frac{135680}{27} \zeta(3) - 8800 \, \zeta(5) + \left(-\frac{91723}{27} \right. \right.$$
$$- \frac{34192}{9} \zeta(3) + 880 \, \zeta(4) + \frac{18400}{9} \zeta(5) \bigg) n_F + \left(\frac{5242}{243} + \frac{800}{9} \zeta(3) \right.$$
$$- \frac{160}{3} \zeta(4) \bigg) n_F^2 + \left(-\frac{332}{243} + \frac{64}{27} \zeta(3) \right) n_F^3 \bigg\} . \qquad (3.15)$$

$$\gamma_4 = \frac{1}{4^5}\left\{\frac{99512327}{162} + \frac{46402466}{243}\,\zeta(3) + 96800\,\zeta(3)^2 - \frac{698126}{9}\,\zeta(4)\right.$$

$$- \frac{231757160}{243}\,\zeta(5) + 242000\,\zeta(6) + 412720\,\zeta(7) + n_f\left[-\frac{150736283}{1458}\right.$$

$$- \frac{12538016}{81}\,\zeta(3) - \frac{75680}{9}\,\zeta(3)^2 + \frac{2038742}{27}\,\zeta(4) + \frac{49876180}{243}\,\zeta(5)$$

$$\left. - \frac{638000}{9}\,\zeta(6) - \frac{1820000}{27}\,\zeta(7)\right] + n_f^2\left[\frac{1320742}{729} + \frac{2010824}{243}\,\zeta(3)\right.$$

$$\left. + \frac{46400}{27}\,\zeta(3)^2 - \frac{166300}{27}\,\zeta(4) - \frac{264040}{81}\,\zeta(5) + \frac{92000}{27}\,\zeta(6)\right]$$

$$+ n_f^3\left[\frac{91865}{1458} + \frac{12848}{81}\,\zeta(3) + \frac{448}{9}\,\zeta(4) - \frac{5120}{27}\,\zeta(5)\right]$$

$$\left. + n_f^4\left[-\frac{260}{243} - \frac{320}{243}\,\zeta(3) + \frac{64}{27}\,\zeta(4)\right]\right\}. \tag{3.16}$$

In order to obtain an expansion of the quark mass in terms of inverse powers of $L \equiv \ln(\mu^2/-q^2)$, needed in applications of FESR, one has to integrate Eq. (3.11) after first solving Eq. (3.1). The power series expansion of the latter is

$$a_s(\mu^2) = \frac{1}{\beta_0 L}\left\{1 - \frac{b_1}{b_0^2}\frac{\ln L}{L} + \frac{1}{b_0^2 L^2}\left[\frac{b_1^2}{b_0^2}\left(\ln^2 L - \ln L - 1\right) + \frac{b_2}{b_0}\right]\right.$$

$$+ \frac{1}{b_0^3 L^3}\left[\frac{b_1^3}{b_0^3}\left(-\ln^3 L + \frac{5}{2}\ln^2 L + 2\ln L - \frac{1}{2}\right)\right.$$

$$\left.\left. - 3\frac{b_1 b_2}{b_0^2}\ln L + \frac{1}{2}\frac{b_3}{b_0}\right] + \mathcal{O}\left(\frac{1}{L^4}\right)\right\}. \tag{3.17}$$

Integrating Eq. (3.11) one finds

$$m_q(\mu^2) = \tilde{m}\left[a_s(\mu^2)\right]^{\gamma_0/\beta_0}\exp\left[c_1 a_s(\mu^2) + c_2\,a_s^2(\mu^2) + c_3\,a_s^3(\mu^2) + \cdots\right], \tag{3.18}$$

where $\tilde{m}$ is an integration constant, and the coefficients c_i are

$$c_1 = \frac{\gamma_1}{b_0} - \frac{\gamma_0\,b_1}{b_0^2}, \tag{3.19}$$

$$c_2 = \frac{1}{2}\left[\frac{\gamma_2}{b_0} - \frac{\gamma_1\,b_1}{b_0^2} + \frac{\gamma_0}{b_0^2}\left(\frac{b_1^2}{b_0} - b_2\right)\right], \tag{3.20}$$

$$c_3 = \frac{1}{3}\left\{\frac{\gamma_0}{b_0^2}\left[\frac{b_1 b_2}{b_0} - \frac{b_1}{b_0}\left(\frac{b_1^2}{b_0} - b_2\right) - b_3\right]\right.$$
$$\left. + \frac{\gamma_1}{b_0^2}\left(\frac{b_1^2}{b_0} - b_2\right) - \frac{b_1 \cdot \gamma_2}{b_0^2} + \frac{\gamma_3}{b_0}\right\}. \tag{3.21}$$

The next step is to expand the overall factor in Eq. (3.18), $\left[a_s(\mu^2)\right]^{\gamma_0/\beta_0}$. For this purpose one invokes the expansion

$$(1+x)^\lambda = 1 + \lambda x + \frac{\lambda(\lambda-1)}{2!}x^2 + \frac{\lambda(\lambda-1)(\lambda-2)}{3!}x^3 + \cdots, \tag{3.22}$$

which leads to

$$[a_s(\mu^2)]^{\gamma_0/b_0} = \left(\frac{1}{b_0 L}\right)^{\gamma_0/b_0}\left(1 + \frac{d_1}{L} + \frac{d_2}{L^2} + \frac{d_3}{L^3}\cdots\right), \tag{3.23}$$

where

$$d_1 = -\frac{\gamma_0 b_1}{b_0^3}\ln L, \tag{3.24}$$

$$d_2 = \frac{\gamma_0}{b_0^3}\left[\frac{b_1^2}{b_0^2}\left(\ln^2 L - \ln L - 1\right) + \frac{b_2}{b_0}\right] + \frac{\gamma_0 b_1^2}{2 b_0^5}\left(\frac{\gamma_0}{b_0} - 1\right)\ln^2 L, \tag{3.25}$$

$$d_3 = \frac{\gamma_0}{b_0^4}\left[\frac{b_1^3}{b_0^3}\left(-\ln^3 L + \frac{5}{2}\ln^2 L + 2\ln L - \frac{1}{2}\right) - 3\frac{b_1 b_2}{b_0^2}\ln L + \frac{1}{2}\frac{b_3}{b_0}\right]$$
$$- \frac{\gamma_0 b_1}{b_0^5}\left(\frac{\gamma_0}{b_0} - 1\right)\left[\frac{b_1^2}{b_0^2}\left(\ln^2 L - \ln L - 1\right) + \frac{b_2}{b_0}\right]\ln L$$
$$- \frac{1}{6}\frac{\gamma_0 b_1^3}{b_0^7}\left(\frac{\gamma_0}{b_0} - 1\right)\left(\frac{\gamma_0}{b_0} - 2\right)\ln^3 L. \tag{3.26}$$

The integration constant $\tilde{m}$ is now redefined as

$$\tilde{m} \rightarrow \hat{m}\,(2 b_0)^{\gamma_0/b_0} \tag{3.27}$$

where $\hat{m}$ is the so-called invariant mass. The quark mass $m_q(\mu^2)$ then becomes

$$m_q(\mu^2) = \frac{\hat{m}_q}{\left(\frac{1}{2}L\right)^{\gamma_0/b_0}} \left\{ 1 + \left(\frac{c_1}{b_0} + d_1\right)\frac{1}{L} + \left[\frac{c_1 e_1}{b_0} + \left(c_2 + \frac{c_1^2}{2}\right)\frac{1}{b_0^2}\right.\right.$$

$$\left. + d_2 + \frac{d_1 c_1}{b_0}\right]\frac{1}{L^2} + \left[\frac{c_1 e_2}{b_0} + \frac{(c_2 + c_1^2/2)}{b_0^2}\left(d_1 - 2\frac{b_1}{b_0^2}\ln L\right)\right.$$

$$\left.\left. + \left(c_3 + c_1 c_2 + \frac{c_1^3}{6}\right)\frac{1}{b_0^3} + d_1 c_1\frac{e_1}{b_0} + d_2\frac{c_1}{b_0} + d_3\right]\frac{1}{L^3} + \mathcal{O}\left(\frac{1}{L^4}\right)\right\}, \quad (3.28)$$

where the coefficients c_i are given in Eqs. (3.19)–(3.21), the d_i in Eqs. (3.24)–(3.26), and

$$e_1 = -\frac{b_1}{b_0^2}\ln L, \qquad (3.29)$$

$$e_2 = \frac{1}{b_0^2}\left[\frac{b_1^2}{b_0^2}\left(\ln^2 L - \ln L - 1\right) + \frac{b_2}{b_0}\right]. \qquad (3.30)$$

It should be noticed that the coefficient γ_4 is now known (see Eq. (3.16)). The reader is encouraged to determine the last term of order $\mathcal{O}\left(\frac{1}{L^4}\right)$ in Eq. (3.28).

Finally, in analogy with Eq. (3.9) the running quark mass can also be expressed in terms of its value at some scale $s = s^*$ according to

$$\bar{m}(s) = \bar{m}(s^*)\left\{1 - a(s^*)\gamma_0\,\eta + \frac{1}{2}a^2(s^*)\eta\left[-2\gamma_1 + \gamma_0\,(\beta_0 + \gamma_0)\,\eta\right]\right.$$

$$- \frac{1}{6}a^3(s^*)\eta\left[6\gamma_2 - 3\left(\beta_1\gamma_0 + 2\,(\beta_0 + \gamma_0)\,\gamma_1\right)\eta + \gamma_0\,(2\beta_0^2 + 3\beta_0\gamma_0 + \gamma_0^2)\,\eta^2\right]$$

$$+ \frac{1}{24}a^4(s^*)\eta\left[-24\gamma_3 + 12(\beta_2\gamma_0 + 2\beta_1\gamma_1 + \gamma_1^2 + 3\beta_0\gamma_2 + 2\gamma_0\gamma_2)\,\eta\right.$$

$$- 4\left(6\beta_0^2\gamma_1 + 3\gamma_0^2\,(\beta_1 + \gamma_1) + \beta_0\gamma_0\,(5\beta_1 + 9\gamma_1)\right)\eta^2 + \gamma_0\,(6\beta_0^3 + 11\beta_0^2\gamma_0$$

$$\left.\left. + 6\beta_0\gamma_0^2 + \gamma_0^3)\,\eta^3\right]\right.$$

$$+ \frac{1}{120} a^5(s^*) \eta \left[- 120 \gamma_4 + \frac{1}{\beta_0} 60 \left(- 7 \beta_1 \beta_2 \gamma_0 + 4 \beta_0^2 \gamma_3 + \beta_0 \left(7 \beta_1 \gamma_0 + \beta_3 \gamma_0 \right. \right. \right.$$

$$\left. + 2 \beta_2 \gamma_1 + 3 \beta_1 \gamma_2 + 2 \gamma_1 \gamma_2 + 2 \gamma_0 \gamma_3 \right) \right) \eta - 20 \left(3 \beta_1^2 \gamma_0 + \beta_1 \left(14 \beta_0 + 9 \gamma_0 \right) \gamma_1 \right.$$

$$\left. + 3 \left(2 \beta_0 + \gamma_0 \right) \left(\beta_2 \gamma_0 + \gamma_1^2 + 2 \beta_0 \gamma_2 + \gamma_0 \gamma_2 \right) \right) \eta^2 + 10 \left(12 \beta_0^3 \gamma_1 + \gamma_0^3 \left(3 \beta_1 + 2 \gamma_1 \right) \right.$$

$$\left. + \beta_0 \gamma_0^2 \left(13 \beta_1 + 12 \gamma_1 \right) + \beta_0^2 \gamma_0 \left(13 \beta_1 + 22 \gamma_1 \right) \right) \eta^3 - \gamma_0 \left(24 \beta_0^4 + 50 \beta_0^3 \gamma_0 \right.$$

$$\left. \left. + 35 \beta_0^2 \gamma_0^2 + 10 \beta_0 \gamma_0^3 + \gamma_0^4 \right) \eta^4 \right] + \mathcal{O}(a^6(s^*)) \right\} \tag{3.31}$$

where $\eta \equiv \ln(s/s^*)$, $a_s \equiv \alpha_s/\pi$, and like Eq. (3.9) this expression makes no use of Λ_{QCD}.

References

1. P.A. Baikov, K.G. Chetyrkin, J.H. Kühn, Phys. Rev. Lett. **118**, 082002 (2017)
2. T. van Ritbergen, J.A.M. Vermaseren, S.A. Larin, Phys. Lett. B **400**, 379 (1997)
3. M. Czakon, Nucl. Phys. B **710**, 485 (2005)
4. A. Pich, Eur. Phys. J. Web Conf. **137**, 01016 (2017)
5. M. Davier, A. Höcker, Z. Zhang, Rev. Mod. Phys. **78**, 1043 (2006)
6. K.G. Patrignani et al., Particle Data Group. Chin. Phys. C **40**, 100001 (2016)
7. P.A. Baikov, K.G. Chetyrkin, J.H. Kühn, J. High Energy Phys. **10**, 076 (2014)
8. T. Luthe, A. Maier, P. Marquard, Y. Schröder, J. High Energy Phys. **1701**, 081 (2017)

Chapter 4
Integration in the Complex s-Plane

In the QCD FESR, Eq. 1.13, the integral around the circle of radius $|s_0|$ can be performed in two ways, named Fixed Order Perturbation Theory (FOPT), and Contour Improved Perturbation Theory (CIPT). In FOPT the strong coupling is *frozen* on the circle, i.e. at $s = |s_0|$, and the renormalization group (RG) improvement, $(\mu^2 = q^2)$, in all logarithms is performed after integration. In CIPT the RG improvement is performed before integration, and $\alpha_s(s)$ is a complex function of s, to be determined by solving the RG equation at each point on the circle. This is done through a single-step numerical contour integration, using as input the strong coupling obtained by solving numerically the RG equation for $\alpha_s(-s)$. This technique achieves a partial resummation of higher order logarithmic integrals, and improves the convergence of the perturbative QCD series. In correlators involving quark masses, e.g. $\psi_5(s)$, Eq. (1.5), CIPT requires the running quark mass also to be integrated around the circle, after computing it at each step by solving the corresponding RG equation.

In general, results from both integration procedures depend on the particular correlation function. In some cases CIPT proves superior to FOPT, but in others there is no major difference. Hence, it has been customary to use both methods in applications.

There are some QCD correlation functions that are either asymptotically constant or that diverge. This would imply subtracted dispersion relations. Alternatively, one could not consider the correlator itself, but rather its (convergent) derivative(s). This situation affects the contour integral in the FESR, Eq. (1.13), if one uses CIPT. Hence, instead of employing the original correlator one can take sufficient derivatives of it. For instance, the pseudoscalar current correlator $\psi_5(q^2)$, Eq. (1.5), diverges asymptotically as $-q^2$. Hence one needs to consider its second derivative. In this case there are some useful identities that are required in the FESR, to wit.

One starts from the identity

$$\oint_{|s_0|} ds\, g(s)\, \psi_5(s) = -\oint_{|s_0|} ds\, [G(s) - G(s_0)]\, \frac{d\psi_5(s)}{ds}, \qquad (4.1)$$

© The Author(s), under exclusive licence to Springer Nature Switzerland AG 2018
C. A. Dominguez, *Quantum Chromodynamics Sum Rules*,
SpringerBriefs in Physics, https://doi.org/10.1007/978-3-319-97722-5_4

where

$$G(s) = \int_0^s ds'\, g(s')\,,\tag{4.2}$$

which follows trivially after substituting $G(s)$, Eq. (4.2), into Eq. (4.1). The analytic function $g(s)$ is some given integration kernel, usually present in the hadronic sector to quench/enhance contributions to FESR. Repeating this procedure leads to a formula involving $\psi_5''(s) \equiv d^2\,\psi_5(s)/ds^2$, which is well behaved at infinity

$$\oint_{|s_0|} ds\, g(s)\, \psi_5(s) = \oint_{|s_0|} ds\, [F(s) - F(s_0)]\, \frac{d^2\psi_5(s)}{ds^2}\,,\tag{4.3}$$

where

$$F(s) = \int_0^s ds'[G(s') - G(s_0)] = \int_0^s ds' \left[\int_0^{s'} ds''\, g(s'') - \int_0^{s_0} ds''\, g(s'') \right].\tag{4.4}$$

Proceeding to the (counter-clockwise) integration in the complex s-plane, the starting point is the definition

$$s = -s_0\, e^{i\phi} = s_0\, e^{i(\phi+\pi)} \equiv s_0\, e^{i\,\alpha}\,,\tag{4.5}$$

so that

$$ds = i\, s_0\, e^{i\,\alpha}\, d\alpha.\tag{4.6}$$

Typical integrals in the pseudoscalar correlator case are

$$
\begin{aligned}
I^a_{N,M}(s_0) &\equiv \frac{1}{2\pi i} \oint_{|s_0|} \frac{ds}{s}\, s^N \left[\frac{\alpha_s(s_0\, e^{i\phi})}{\pi} \right]^M \\
&= \frac{i}{2\pi i}\, (-s_0)^N \left[\int_{-\pi}^0 d\phi\, e^{iN\phi} \left(\frac{\alpha_s}{\pi} \right)^M + \int_0^\pi d\phi\, e^{iN\phi} \left(\frac{\alpha_s}{\pi} \right)^M \right] \\
&= \frac{1}{\pi}\, (-s_0)^N\, Re \int_0^\pi d\phi\, e^{iN\phi} \left[\frac{\alpha_s(s_0\, e^{i\phi})}{\pi} \right]^M\,,
\end{aligned}\tag{4.7}
$$

$$
\begin{aligned}
I^b_{N,M}(s_0) &\equiv \frac{1}{2\pi i} \int_{-\pi}^0 d\phi \left[\frac{\alpha_s(s_0\, e^{i\phi})}{\pi} \right]^M + \frac{1}{2\pi i} \int_0^\pi d\phi \left[\frac{\alpha_s(s_0\, e^{i\phi})}{\pi} \right]^M\,, \\
&= \frac{1}{\pi}\, Re \cdot \int_0^\pi d\phi \left[\frac{\alpha_s(s_0\, e^{i\phi})}{\pi} \right]^M\,,
\end{aligned}\tag{4.8}
$$

where the angular integration is performed counter-close-wise, and the last step in Eqs. (4.7)–(4.8) follows from the requirement that the integrals be real.

Depending on the current correlator, not only the strong coupling, but also the running quark mass needs to be integrated around the circle on the complex s-plane. Rearranging Eq. (3.11) one finds

$$\frac{dm_q}{m_q} = \frac{\gamma(a_s)}{\beta(a_s)}\,da_s \tag{4.9}$$

which is the equation to be integrated

$$\int_{m(\mu)}^{m(q)} \frac{dm}{m} = \ln\left[\frac{m(q)}{m(\mu)}\right] = \int_{a_s(\mu)}^{a_s(q)} \frac{\gamma(a_s)}{\beta(a_s)}\,da_s. \tag{4.10}$$

If one were to expand the ratio

$$R(a_s) \equiv \frac{\gamma(a_s)}{\beta(a_s)}, \tag{4.11}$$

one looses information on two orders in a_s. For instance, to keep terms of order $\mathcal{O}(a_s^3)$ one needs information on γ_5 and β_5, currently unknown. Hence this is not a good idea. The way out is rather simple, to wit. Starting from the RGE, Eq. 4.9, and substituting

$$da_s(x) = i\,\beta(a_s)\,dx, \tag{4.12}$$

gives

$$m_q(x) = m_q(x_0)\,\exp\left[i\int_{x_0}^{x}\gamma[a_s(x')]\,dx'\right], \tag{4.13}$$

which will make use of full information up to $\mathcal{O}(a_s^4)$. In practice, using CIPT one needs to integrate numerically Eq. (4.13), with $x_0 = 0$ and $m_q(x_0) = m_q(s_0)$ so that

$$m_q(x) = m_q(s_0)\,exp\left[-i\int_0^x dx'\sum_J \gamma_J\,[a_s(x')]^J\right]. \tag{4.14}$$

To complete this section let us consider the numerical integration of the strong coupling and the quark mass in the complex s-plane. The relation between s and the angle $\phi \equiv x \in (-\pi, \pi)$ is given in Eq. (4.5), so that the equation to be solved numerically is

$$\frac{da_s(x)}{dx} = -i\sum_{N=0}b_N\left[a_s(x)\right]^{N+2}, \tag{4.15}$$

with the input $a_s(x = 0) = a_s\left(s_0\,e^{i(x+\pi)}\right)|_{x=0} = a_s(-s_0)$. This differential equation is easily solved using Euler's method, to wit.

Given a function $f(x, y)$, with $y = y(x)$, and $dy(x)/dx = f(x, y)$, with initial condition $y(0)$, then

$$y_{(i+1)} = y_i + h\,f\left[x_i + \frac{h}{2},\; y_i + \frac{h}{2}\,f(x_i, y_i)\right], \tag{4.16}$$

which converges after a few steps. As an example, consider the function

$$\frac{dy(x)}{dx} = -2x\,y^2(x) \equiv f(x, y),$$
(4.17)

with $x_1 = 0$, and $y(0) = y(1) = 1$. The exact analytic solution is

$$y(x) = \frac{1}{1 + x^2}.$$
(4.18)

Choosing the point $x = 1$, gives $y(1) = 0.5$. Proceeding to iterate Eq. (4.16) with $h = 0.2$ and

$$y_1 = 1 \;(input),$$
(4.19)

gives

$$y_2 = 1 + 0.2\, f\left[\frac{h}{2}, \; 1 + \frac{h}{2} * 0\right] = 1 + 0.2\, f(0.1, 1) = 0.96,$$
(4.20)

$$y_3 = 0.96 + 0.2\, f\left[0.2 + \frac{h}{2}, \; 0.96 + \frac{h}{2}\, f(0.2, 0.96)\right] = 0.8577,$$
(4.21)

$$\cdots \; y_5 = 0.498.$$
(4.22)

In the case in which the function depends only on y,

$$\frac{dy(x)}{dx} = f(y),$$
(4.23)

the solution is the iteration of

$$y_{(i+1)} = y_i + h\, f\left[y_i + \frac{h}{2}\, f(y_i)\right].$$
(4.24)

Chapter 5
Determination of the QCD Strong Coupling

The idea that hadronic decays of the τ-lepton could provide an ideal laboratory for studying hadronic weak currents at low and intermediate energy was first proposed in [1]. In fact, in the light-quark sector the QCD strong coupling can now be determined with high accuracy from experimental data on τ-lepton decay [2, 3]. After extrapolation of this value to the Z-boson mass there is excellent agreement with independent determinations in that region.

The ratio R in τ-decay is defined as

$$R_\tau = \frac{\Gamma(\tau^- \to \nu_\tau \text{ hadrons})}{\Gamma(\tau^- \to \nu_\tau e^- \bar{\nu}_e)} \tag{5.1}$$

which can be written in terms of QCD vector and axial-vector current correlators as

$$R_\tau = 12\pi S_{EW} \int_0^{m_\tau^2} \frac{ds}{m_\tau^2} \left(1 - \frac{s}{m_\tau^2}\right)^2 \left[\left(1 + 2\frac{s}{m_\tau^2}\right)\text{Im}\Pi^{(1)}(s) + \text{Im}\Pi^{(0)}(s)\right], \tag{5.2}$$

where $S_{EW} = 1.0201 \pm 0.0003$ is the electro-weak correction [4], and

$$\Pi^{(J)}(s) = \sum_{q=d,s} |V_{uq}|^2 \left(\Pi_{uq,V}^{(J)}(s) + \Pi_{uq,A}^{(J)}(s)\right), \tag{5.3}$$

where V_{uq} is the Cabibbo-Kobayashi-Maskawa matrix element, $J = 0, 1$, and $\Pi_{uq,V}^{(J)}(s)$ and $\Pi_{uq,A}^{(J)}(s)$ are the vector and axial-vector light-quark current correlators

$$i \int d^4x\, e^{iqx} \langle 0|T(J_\mu(x) J_\nu(0))\rangle = (-g_{\mu\nu}q^2 + q_\mu q_\nu)\Pi_{V,A}^{(1)}(q^2) + q_\mu q_\nu \Pi_{V,A}^{(0)}(q^2), \tag{5.4}$$

where a simplified notation is implicit, and $J_\mu(x)$ stands for the vector, $V_\mu(x)$, or the axial-vector, $A_\mu(x)$, currents. In the sequel we concentrate on the non-strange current correlators, in which case there is no longitudinal term in the vector channel,

© The Author(s), under exclusive licence to Springer Nature Switzerland AG 2018   25
C. A. Dominguez, *Quantum Chromodynamics Sum Rules*,
SpringerBriefs in Physics, https://doi.org/10.1007/978-3-319-97722-5_5

i.e. $\Pi_V^{(0)}(q^2)$ is absent in Eq. (5.4), while $\Pi_A^{(0)}(q^2)$ corresponds to the pion pole. In perturbative QCD (PQCD), and for massless quarks, the vector correlator is identical to the axial-vector and is given by

$$
\begin{aligned}
-4\pi^2\,\Pi_{V,A}(q^2) = {}& L + L\,a_s + a_s^2\left(L\,k_2 - \frac{1}{2}L^2\,b_0\right) \\
& + a_s^3\left[\frac{1}{3}L^3\,b_0^2 + L\,k_3 - L^2\left(\frac{1}{2}\,b_1 + b_0\,k_2\right)\right] \\
& + a_s^4\left[L\,k_4 - \frac{1}{4}L^4 b_0^3 - L^2\left(\frac{1}{2}b_2 + \frac{3}{2}\,b_0\,k_3 + b_1\,k_2\right)\right. \\
& \left. + \frac{1}{6}\,L^3\,b_0\,(5\,b_1 + 6\,b_0\,k_2)\right] \\
& + a_s^5\left[L\,k_5 - L^2\left(\frac{1}{2}\,b_3 + 2\,b_0\,k_4 + \frac{3}{2}\,b_1\,k_3 + b_2\,k_2\right)\right. \\
& + L^3\left(2\,k_3\,b_0^2 + \frac{7}{3}\,k_2\,b_0\,b_1 + b_2\,b_0 + \frac{1}{2}\,b_1^2\right) \\
& \left. - \frac{1}{12}L^4\,b_0^2\left(13\,b_1 + 12\,b_0\,k_2 + \frac{1}{5}L^5\,b_0^4\right)\right],
\end{aligned}
\tag{5.5}
$$

where $a_s \equiv \alpha_s/\pi$, $q^2 < 0$, $L \equiv \ln(-q^2/\mu^2)$, b_i are given in Eqs. (3.3)–(3.7), $k_1 = 1$, $k_2 = 1.63982$, $k_3 = 6.37101$, and $k_4 = 49.076$.

An alternative procedure involves the Adler function, $D(q^2)$ rather than $\Pi_{V,A}(q^2)$ itself [2, 3]. The Adler function, $D(q^2)$, is defined as

$$
D(q^2) = -q^2\,\frac{d}{dq^2}\,\Pi_{V,A}(q^2)\,.
\tag{5.6}
$$

There is a renormalization group constraint involving $D(q^2)$

$$
\mu^2\,\frac{d}{d\mu^2}\,D[\ln(-q^2/\mu^2), a_s(\mu^2)] = 0\,,
\tag{5.7}
$$

where $q^2 < 0$. For details on this approach see [2, 3].

Turning to Eq. (5.2), the term proportional to s in R_τ, Eq. (5.2) vanishes. Hence, there is no contribution from the dimension-four gluon condensate, and R_τ is essentially given by perturbative QCD, and the strong coupling a_s becomes a function of R_τ.

The latest determination of $\alpha_s(M_\tau)$ in this framework is [3]

$$\alpha_s(M_\tau^2) = 0.328 \pm 0.013, \qquad (5.8)$$

which gives $\alpha_s(M_Z^2) = 0.1197 \pm 0.0015$, in good agreement with the electroweak precision fit [4] $\alpha_s(M_Z^2) = 0.1196 \pm 0.0030$, and the world average value $\alpha_s(M_Z^2) = 0.1181 \pm 0.0011$ [5].

It is unfortunate that the τ-lepton is not heavier, as it would have allowed for precise information on the onset of QCD, as well as on potential duality violations. As it stands, there is only a hint of PQCD starting at, and near, the end-point $s \simeq M_\tau^2$. However, it has been shown [6] that it is possible to use a suitable QCD-FESR to effectively extend the kinematical range beyond the end-point of the data up to $s \simeq 10\,\text{GeV}^2$. By construction, this FESR suppresses the hadronic contribution along the real s-axis in the complex s-plane, beyond the kinematical end-point of τ-decay, $s_1 = M_\tau^2$, where there is no longer experimental data.

The starting point is the axial-vector current correlator

$$i \int d^4 x\, e^{iqx} \langle 0|T(A_\mu(x) A_\nu(0))|0\rangle = (-g_{\mu\nu} q^2 + q_\mu q_\nu) \Pi_A^{(1)}(q^2) + q_\mu q_\nu \Pi_A^{(0)}(q^2),$$

$$(5.9)$$

where $A_\mu(x) = \bar{d}(x)\gamma_\mu\gamma_5 u(x)$ is the light-quark axial-vector current. Next, one invokes the general FESR, Eq. (1.13), which after using the OPE, Eq. (2.2), it becomes

$$(-)^N C_{2N+2}\langle \mathcal{O}_{2N+2}\rangle = 4\pi^2 \int_0^{s_0} ds\, s^N \frac{1}{\pi}\text{Im}\Pi(s)|_{HAD} - s_0^{N+1} M_{2N+2}(s_0),$$

$$(5.10)$$

where the dimensionless perturbative QCD (PQCD) moments, $M_{2N+2}(s_0)$, are

$$M_{2N+2}(s_0) = \frac{4\pi^2}{s_0^{(N+1)}} \int_0^{s_0} ds\, s^N \frac{1}{\pi}\, Im\, \Pi(s)|_{PQCD}. \qquad (5.11)$$

For $N = 0$, $C_2\langle \mathcal{O}_2\rangle = 0$ (no condensate of dimension $d = 2$), and

$$\text{Im}\,\Pi(s)|_{HAD} = 2 f_\pi^2, \qquad (5.12)$$

the FESR Eq. (5.10) gives f_π as a function of s_0, shown in Fig. 5.1 for $\alpha_s(M_\tau^2) = 0.335$. Except possibly near the end-point, the prediction is not good. In fact, it is well known that the Weinberg sum rules, involving the difference of the vector and the axial-vector correlators are not well saturated by the ALEPH data, unless one introduces pinched kernels in the FESR [7, 8].

In view of this situation it was proposed in [6] to introduce a polynomial integration kernel in the FESR, Eq. (5.10), tuned to eliminate the (unknown) hadronic contribution in the interval $s_1 - s_0$, where $s_1 \simeq M_\tau^2$, and $s_0 \gg M_\tau^2$. The optimal integration kernel turns out to be the simple linear function

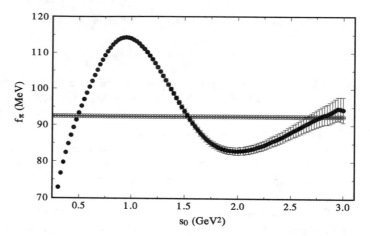

Fig. 5.1 Results for f_π from the FESR Eq. (5.10) as a function of s_0. Straight line is the experimental value of f_π

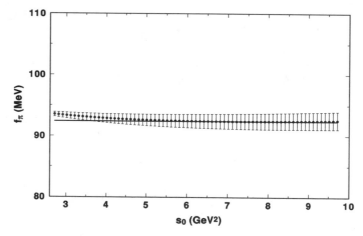

Fig. 5.2 Results for $F(s_0) \equiv f_\pi^2$ from the FESR Eq. (5.15) in the axial-vector channel as a function of s_0. Straight line is the experimental value of f_π

$$P(s) = 1 - \frac{2s}{s_0 + s_1}, \qquad (5.13)$$

with the requirement

$$C \times \int_{s_1}^{s_0} P(s)\, ds = 0, \qquad (5.14)$$

where C is a constant. With this choice the FESR, Eq. (5.10), becomes

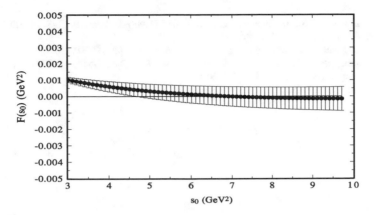

Fig. 5.3 Results for $F(s_0) = 0$ from the FESR Eq. (5.15) in the vector channel as a function of s_0. Straight line is the left-hand-side of Eq. (5.15)

$$F(s_0) = -\int_0^{s_1} ds\, P(s)\frac{1}{\pi}\, Im\,\Pi(s)|_{DATA} + \frac{s_0}{4\pi^2}\left[M_2(s_0) - \frac{2s_0}{s_0 + s_1}M_4(s_0)\right]$$

$$+ \frac{1}{4\pi^2}\left[C_2\langle\mathcal{O}_2\rangle + \frac{2}{s_0 + s_1}C_4\langle\mathcal{O}_4\rangle\right] + \Delta(s_0), \qquad (5.15)$$

where $F(s_0) \equiv 2f_\pi^2$ in the axial-vector channel, and $F(s_0) = 0$ in the vector channel. The pion pole has been separated from the data, and the chiral limit is understood. The term $\Delta(s_0)$ represents the uncertainty due to assuming the unknown data to be constant in the interval $s_1 - s_0$. If the onset of PQCD would be at $s = s_1$ then the data would follow the logarithmic fall-off of PQCD. A detailed discussion of the caveats of this procedure is given in the original paper [6].

In Fig. 5.2 we show the results for f_π from the FESR, Eq. (5.15), to be compared with Fig. 5.1. The same procedure can be followed in the vector current channel, where now $F(s_0) = 0$, as there is no (hadronically) stable scalar analogue of the pion. The result for $F(s_0)$ is shown in Fig. 5.3, exhibiting very good agreement in a wide range of s_0. In the vector channel there is independent data from the process e^+e^- annihilation into hadrons. This allows for an estimate of the uncertainty $\Delta(s_0)$ in Eq. (5.15). The result is $\Delta(s_0)|_V \simeq (10^{-5} - 10^{-4})\,\text{GeV}^2$, i.e. 1–2 orders of magnitude smaller than $F(s_0)$ (see Fig. 5.3), thus supporting the method. For more details see [6].

References

1. K. Schilcher, M.D. Tran, Phys. Rev. D **29**, 570 (1984)
2. M. Davier, A. Höcker, Z. Zhang, Rev. Mod. Phys. **78**, 1043 (2006)
3. A. Pich, A. Rodriguez-Sanchez, Phys. Rev. D **94**, 034027 (2016)

4. W.J. Marciano, A. Sirlin, Phys. Rev. Lett. **61**, 1815 (1988); E. Braaten, C.-S. Li, Phys. Rev. D **42**, 3888 (1990); J. Erler, Rev. Mex. Fis. **50**, 200 (2004)
5. K.G. Patrignani et al., Particle data group. Chinese Phys. C **40**, 100001 (2016)
6. C.A. Dominguez, N.F. Nasrallah, K. Schilcher, Phys. Rev. D **80**, 054014 (2009)
7. K. Maltman, Phys. Lett. B **440**, 367 (1998)
8. C.A. Dominguez, K. Schilcher, Phys. Lett. B **581**, 193 (2004)

Chapter 6
Hadronic Spectral Functions

Chapters 2–4 have dealt with the main issues related to the QCD correlator entering the contour integral in the FESR, Eq. (1.13). The next step is to consider the line integral along the real s-axis. In the interval $s = s_{th} - s_0$, the line integral requires hadronic information. This could be experimental data, a parametrization of the spectral function in terms of hadronic degrees of freedom, or some combination of both. Regarding the data, the main sources are $e^+e^- \rightarrow$ hadrons, determining the vector spectral function in the various energy regions, and hadronic decays of the τ-lepton, determining both the vector and axial-vector spectral functions below the kinematical end-point $s = M_\tau^2$.

Starting with the pseudoscalar correlator in the up- down-quark sector, $\psi_5(q^2)$, defined in Eq. (1.15), the hadronic spectral function involves the pion pole term followed by the resonance contribution, i.e.

$$\frac{1}{\pi} \text{Im } \psi_5(s) = 2 f_\pi^2 M_\pi^4 \delta(s - M_\pi^2) + \frac{1}{\pi} \text{Im } \psi_5(s)|_{\text{RES}} , \qquad (6.1)$$

where Im $\psi_5(s)|_{\text{RES}}$, involves the radial excitations of the pion, $\pi(1300)$ and $\pi(1800)$, with masses and widths known from experiment [1]. However, the resonance spectral function itself is unknown experimentally, thus leading to a serious systematic uncertainty. For instance, non-resonant background and resonance interference, are sources of this uncertainty. In an attempt to reduce this systematic uncertainty, it was first proposed in [2] to normalize this spectral function at threshold using the chiral perturbation theory (CHPT) constraint [3]

$$\frac{1}{\pi} \text{Im } \psi_5(s)|_{\pi\pi\pi} = \theta(s) \frac{1}{3} \frac{M_\pi^4}{f_\pi^2} \frac{1}{2^8 \pi^4} s . \qquad (6.2)$$

With this threshold normalization the spectral function involving two pionic radial excitations becomes

© The Author(s), under exclusive licence to Springer Nature Switzerland AG 2018
C. A. Dominguez, *Quantum Chromodynamics Sum Rules*,
SpringerBriefs in Physics, https://doi.org/10.1007/978-3-319-97722-5_6

$$\frac{1}{\pi} \, Im \, \psi_5(s)|_{\text{RES}} \; = Im \, \psi_5(s)|_{\pi\pi\pi} \frac{[BW_1(s) + \kappa \, BW_2(s)]}{(1+\kappa)} \; , \tag{6.3}$$

where $BW_1(0) = BW_2(0) = 1$, with

$$BW_i(s) = \frac{M_i^2(M_i^2 + \Gamma_i^2)}{(s - M_i^2)^2 + M_i^2 \Gamma_i^2} \quad (i = 1, 2) \, , \tag{6.4}$$

and κ is a free parameter controlling the relative weight of the resonances.

A threshold expression beyond the chiral limit was obtained in [4], corrected for misprints in [5]

$$\frac{1}{\pi} \, Im \, \psi_5(s)|_{\pi\pi\pi} = \theta(s) \, \frac{1}{9} \frac{M_\pi^4}{f_\pi^2} \frac{1}{2^8 \, \pi^4} \, I_{PS}(s) \; . \tag{6.5}$$

where the phase-space integral $I_{PS}(s)$ is given by

$$
\begin{aligned}
I_{PS}(s) \; = \; & \int_{4M_\pi^2}^{(\sqrt{s}-M_\pi)^2} du \, \sqrt{1 - \frac{4M_\pi^2}{u}} \, \lambda^{1/2}(1, u/s, M_\pi^2/s) \left\{ 5 + \frac{1}{2} \frac{1}{(s - M_\pi^2)^2} \right. \\
& \times \left[(s - 3u + 3M_\pi^2)^2 + 3 \, \lambda(s, u, M_\pi^2) \left(1 - \frac{4 M_\pi^2}{u} \right) + 20 \, M_\pi^4 \right] \\
& + \frac{1}{(s - M_\pi^2)} \left[3(u - M_\pi^2) - s + 9M_\pi^2 \right] \left. \right\} \, ,
\end{aligned}
\tag{6.6}
$$

where

$$\lambda(1, u/s, M_\pi^2/s) \equiv \left[1 - \frac{\left(\sqrt{u} + M_\pi \right)^2}{s} \right] \left[1 - \frac{\left(\sqrt{u} - M_\pi \right)^2}{s} \right] , \tag{6.7}$$

$$\lambda(s, u, M_\pi^2) \equiv \left[s - \left(\sqrt{u} + M_\pi \right)^2 \right] \left[s - \left(\sqrt{u} - M_\pi \right)^2 \right] . \tag{6.8}$$

In the chiral limit the phase space integral $I_{PS}(s)$ reduces to the simple expression

$$\lim_{M_\pi^2 \to 0} I_{PS}(s) = 3s \, , \tag{6.9}$$

leading to Eq. (6.2), which is an excellent approximation.

This procedure has been extended to the $SU(3) \times SU(3)$ case [6], relevant to the current correlator, Eq. (1.15), with

$$j_5(x) = (m_s + m_{ud}) : \bar{s}(x) i \, \gamma_5 u(x) : , \tag{6.10}$$

with m_s the strange-quark mass, and $m_{ud} = (m_u + m_d)/2$. The result is [6]

$$\frac{1}{\pi} \, Im \, \psi_5(s)|_{K\pi\pi} = \theta(s - M_K^2) \, \frac{M_K^4}{2f_\pi^2} \, \frac{3}{2^8 \, \pi^4} \, \frac{I(s)}{s(M_K^2 - s)^2} \,, \qquad (6.11)$$

where

$$
\begin{aligned}
I(s) \;=\; & \int_{M_K^2}^{s} \frac{du}{u} (u - M_K^2)(s - u) \left\{ (M_K^2 - s) \left[u - \frac{(s + M_K^2)}{2} \right] \right. \\
& - \frac{1}{8u}(u^2 - M_K^4)(s - u) + \left. \frac{3}{4}(u - M_K^2)^2 \, |F_{K^*}(u)|^2 \right\} ,
\end{aligned}
\qquad (6.12)
$$

and

$$|F_{K^*}(u)|^2 = \frac{(M_{K^*}^2 - M_K^2)^2 + M_{K^*}^2 \, \Gamma_{K^*}^2}{(M_{K^*}^2 - u)^2 + M_{K^*}^2 \Gamma_{K^*}^2} \,, \qquad (6.13)$$

where $K^* \equiv K^*(892)$ is the strange vector $K\pi$ resonance. Notice that the version of Eq. (6.11) in [6] has misprints.

In order to reduce the model dependency of the hadronic resonance parametrization it has been proposed [8–11] to choose an integration kernel in the FESR, Eq. (1.13), such that it vanishes at the peak of each resonance

$$\Delta_5(s) = 1 - a_0 \, s - a_1 \, s^2 \,, \qquad (6.14)$$

so that $\Delta_5(M_1^2) = \Delta_5(M_2^2) = 0$. which fixes a_0 and a_1. This simple kernel does achieve a substantial reduction of the systematic uncertainty arising in the hadronic sector, as discussed later in Chap. 9. Ultimately, the optimal choice of integration kernel is most likely application dependent.

Other important hadronic spectral functions needed in applications of QCD sum rules are the axial-vector and the vector spectral functions in the light-quark sector. Starting with the axial-vector case, the spectral function was defined in Eq. (5.4). Concentrating on the longitudinal term, $\Pi_A^{(0)}(q^2)$, the time ordered product of the axial-vector currents can be written as the Fourier transform

$$\langle 0|T(A_\mu(x)A_\nu^\dagger(0))|0\rangle = \int \frac{d^3p}{2 \, p_0 \, (2\pi)^3} \, e^{-ip\cdot x} \, \langle 0|A_\mu(0)|\pi(p)\rangle\langle\pi(p)|A_\nu^\dagger(0)|0\rangle \,, \qquad (6.15)$$

where

$$\langle 0|A_\mu(x)|\pi(p)\rangle|_{x=0} = i \sqrt{2} \, f_\pi \, p_\mu \,, \qquad (6.16)$$

with $f_\pi = 92.28 \pm 0.07\,\text{MeV}$ the (charged) pion decay constant [1]. Substituting Eqs. (6.16) into (6.15), followed by the substitution of Eqs. (6.15) into (5.4) gives

$$
\Pi_{\mu\nu}^{AA} = 2\,f_\pi^2\,i\,\int d^4 x\,e^{iqx}\,\int \frac{d^3 p}{2\,p_0\,(2\pi)^3}\,e^{-ip\cdot x}\,\Delta_{\mu\nu}(p)
$$
$$
= 2 f_\pi^2 \int d^4 x\,e^{iqx}\,\Delta_{\mu\nu}(x) = 2 f_\pi^2 \Delta_{\mu\nu}(q^2) = \left[\frac{2 f_\pi^2}{q^2 - M_\pi^2 + i\epsilon}\right] q_\mu q_\nu, \quad (6.17)
$$

which after confronting with Eq. (5.4) leads to

$$
\Pi_A^{(0)}(q^2) = \frac{2\,f_\pi^2}{q^2 - M_\pi^2 + i\epsilon}, \qquad (6.18)
$$

and

$$
\frac{1}{\pi}\,\text{Im}\,\Pi_A^{(0)}(q^2) = 2\,f_\pi^2\,\delta(q^2 - M_\pi^2). \qquad (6.19)
$$

Continuing with the axial-vector channel, the other correlation function is the transverse $\Pi_A^{(1)}(q^2)$, involving the axial-vector three-pion resonance $a_1(1260)$, of mass $M_{a_1} = 1230 \pm 40\,\text{MeV}$, and broad width $\Gamma = 250 - 600\,\text{MeV}$ [1]. The main decay mode is into $\rho\pi$, followed by the decay of the rho-meson into two pions. Given the width of the a_1 a Breit-Wigner parametrization is out of the question. Instead a fit to the ALEPH data in the axial-vector channel [7], from threshold up to $s = 2.0\,\text{GeV}^2$ is given as [12]

$$
\frac{1}{\pi}\text{Im}\Pi_A(s)|_{a_1} = C\,f_{a_1} \exp\left[-\left(\frac{s - M_{a_1}^2}{\Gamma_{a_1}^2}\right)^2\right]
$$
$$
(0 \le s \le 1.2\,\text{GeV}^2)\,, \qquad (6.20)
$$

$$
\frac{1}{\pi}\text{Im}\Pi_A(s)|_{a_1} = C\,f_{a_1} \exp\left[-\left(\frac{1.2\,\text{GeV}^2 - M_{a_1}^2}{\Gamma_{a_1}^2}\right)^2\right]
$$
$$
(1.2\,\text{GeV}^2 \le s \le 1.45\,\text{GeV}^2)\,, \qquad (6.21)
$$

$$\frac{1}{\pi} \text{Im}\Pi_A(s)|_{a_1} = C\, f_{a_1} \exp\left[-\left(\frac{s - M_{a_1}^2}{\Gamma_{a_1}^2}\right)^2\right]$$

$$(1.45\,\text{GeV}^2 \le s \le M_\tau^2),\qquad\qquad (6.22)$$

where $M_{a_1} = 1.0891$ GeV, $\Gamma_{a_1} = 568.78$ GeV, $C = 0.662$ and $f_{a_1} = 0.073$. This spectral function is shown in Fig. 6.1 together with the ALEPH data [7].

Turning to the vector channel, $\Pi_V^{(1)}(q^2)$ in Eq. (5.4) with $J_\mu(x) \equiv V_\mu(x)$ the hadronic representation is obtained as follows. The time ordered product in the hadronic representation, due to the ρ-meson, is

$$\langle 0|T(V_\mu(x)V_\nu^\dagger(0))|0\rangle|_{HAD} = \sum_s \int_p e^{-ip\cdot x}\, \langle 0|V_\mu(x)|\rho(p,s)\rangle\langle\rho(p,s)|V_\nu^\dagger(0)|0\rangle,$$

$$(6.23)$$

where a shortened notation is to be understood, and

$$\langle 0|V_\mu(0)|\rho(p,s)\rangle = \frac{M_\rho^2}{f_\rho}\,\epsilon_\mu, \quad \langle\rho(p,s)|V_\nu^\dagger(0)|0\rangle = \frac{M_\rho^2}{f_\rho}\,\epsilon_\nu,\qquad (6.24)$$

where $f_\rho = 4.97 \pm 0.07$ from the leptonic decay of the rho-meson [1]. Recalling

$$\sum_s \epsilon_\mu(p,s)\,\epsilon_\nu(p,s) = -g_{\mu\nu} + \frac{p_\mu p_\nu}{M_\rho^2} \equiv \Delta_{\mu\nu}(p),\qquad (6.25)$$

one has for the vector correlator

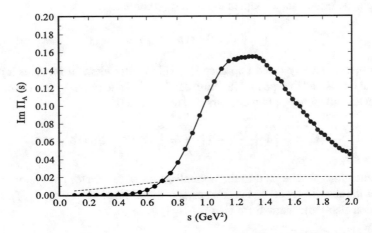

Fig. 6.1 Solid curve is the axial-vector (a_1-resonance) spectral function, Eqs. (6.20)–(6.22), fitted to the ALEPH data [7], shown with error bars the size of the data points. Dotted line is background, to be ignored here

$$\Pi_{\mu\nu}^{V}(q^2) = i \int d^4x \, e^{iq\cdot x} \frac{M_\rho^4}{f_\rho^2} \int \frac{d^3p}{2p_0(2\pi)^3} e^{-ip\cdot x} \Delta_{\mu\nu}(p) \tag{6.26}$$

leading to

$$\Pi_{\mu\nu}^{V}(q^2) = (q_\mu q_\nu - q^2 \, g_{\mu\nu}) \, \Pi_V^{(1)}(q^2) \tag{6.27}$$

where

$$\Pi_V^{(1)}(q^2) = \frac{M_\rho^2}{f_\rho^2} \frac{1}{(q^2 - M_\rho^2)} \,. \tag{6.28}$$

Given the relatively narrow width of the ρ-meson, it is customary to approximate the imaginary part of $\Pi_V^{(1)}(q^2)$ by a delta-function

$$\mathrm{Im}\,\Pi_V^{(1)}(q^2) = \frac{M_\rho^2}{f_\rho^2} \pi \, \delta(q^2 - M_\rho^2) \,. \tag{6.29}$$

If more accuracy is needed, the next level is a Breit-Wigner parametrization

$$\frac{1}{\pi} \mathrm{Im}\Pi_V^{(1)}(s) = \frac{1}{\pi} \frac{1}{f_\rho^2} \frac{M_\rho^3 \, \Gamma_\rho}{\left(s - M_\rho^2\right)^2 + M_\rho^2 \, \Gamma_\rho^2} \,, \tag{6.30}$$

where $s \equiv q^2 > 0$, and a normalization such that this function reduces to Eq. (6.29) in the zero-width approximation. This result is not strictly valid near threshold, as the leading two-pion state coupled to the vector current,

$$\langle 0 | V_\mu(0) | \pi(p_1) \pi(p_2) \rangle \tag{6.31}$$

has not been explicitly taken into account. This matrix element involves a p-wave threshold factor, and the spectral function can be related to the ratio $R(s)$ in electron-positron annihilation. For two flavours this relation is

$$R(s) = \frac{5}{3} \left(1 + \frac{\alpha_s}{\pi} + \cdots \right) = \frac{5}{3} (8\,\pi^2) \frac{1}{\pi} \mathrm{Im}\,\Pi_V^{(1)}(s) \,, \tag{6.32}$$

where attention must be paid to the normalization of the vector correlator, as $R(s)$ involves the electrically neutral vector current. In the threshold region $R(s)$ is related to the pion electromagnetic form factor, $F_\pi(s)$, through

$$R\big(e^+ e^- \to \pi^+ \pi^-\big) = \frac{1}{4} \left(1 - \frac{4 m_\pi^2}{s} \right)^{\frac{3}{2}} |F_\pi(s)|^2 \,. \tag{6.33}$$

Substituting this expression into Eq. (6.32) gives a relation between the spectral function and the pion form factor

$$\text{Im} \, \Pi_V^{(1)}(s) = \frac{3}{160 \, \pi} \left(1 - \frac{4 \, m_\pi^2}{s} \right)^{\frac{3}{2}} |F_\pi(s)|^2 \,, \qquad (6.34)$$

which is a more accurate expression than the simple Breit-Wigner Eq. (6.30). The pion form factor in this expression can be obtained from e^+e^- data, or from τ-lepton decay data in the vector channel [7]. Attention must be paid to the difference in the currents involved in these two processes, i.e. the former involves electrically neutral currents, while the latter involves electrically charged currents. The normalization of these vector current correlators then differs by a factor-two!

References

1. K.G. Patrignani et al., Particle data group. Chin. Phys. C **40**, 100001 (2016)
2. C. A. Dominguez, Z. Phys. C **26**, 269 (1984)
3. H. Pagels, A. Zepeda, Phys. Rev. D **5**, 3262 (1972)
4. C.A. Dominguez, E. de Rafael, Ann. Phys. (NY) **174**, 372 (1987)
5. J. Bijnens, J. Prades, E. de Rafael, Phys. Lett. B **226** (1995)
6. C.A. Dominguez, L. Pirovano, K. Schilcher, Phys. Lett. B **425**, 193 (1998)
7. M. Davier, A. Höcker, Z. Zhang, Rev. Mod. Phys. **78**, 1043 (2006)
8. J. Bordes, C.A. Dominguez, P. Moodley, J. Penarrocha, K. Schilcher, J. High Energy Phys. **1005**, 064 (2010)
9. C.A. Dominguez, N.F. Nasrallah, R. Röntsch, K. Schilcher, J. High Energy Phys. **05**, 020 (2008)
10. C.A. Dominguez, N.F. Nasrallah, R. Röntsch, K. Schilcher, J. High Energy Phys. **02**, 072 (2008)
11. C.A. Dominguez, N.F. Nasrallah, R. Röntsch, K. Schilcher, Phys. Rev. D **79**, 014009 (2009)
12. A. Ayala, C.A. Dominguez, M. Loewe, Y. Zhang, Phys. Rev. D **90**, 034012 (2014)

Chapter 7
QCD Chiral Sum Rules

This section deals with a few sum rules first obtained as consequences of chiral symmetry [1], i.e. long before QCD. First, and foremost, the two Weinberg sum rules (WSR) [2], derived from current algebra and chiral $SU(2) \times SU(2)$ symmetry are given by

$$W_1 \equiv \int_0^\infty ds \, \frac{1}{\pi} \left[\text{Im} \Pi_V(s) - \text{Im} \Pi_A(s) \right] = 2 f_\pi^2 , \qquad (7.1)$$

$$W_2 \equiv \int_0^\infty ds \, s \, \frac{1}{\pi} \left[\text{Im} \Pi_V(s) - \text{Im} \Pi_A(s) \right] = 0 , \qquad (7.2)$$

where

$$\Pi_{\mu\nu}^{VV}(q^2) = i \int d^4 x \, e^{iqx} < 0 | T(V_\mu(x) V_\nu^\dagger(0)) | 0 >$$
$$= (-g_{\mu\nu} q^2 + q_\mu q_\nu) \, \Pi_V(q^2) , \qquad (7.3)$$

$$\Pi_{\mu\nu}^{AA}(q^2) = i \int d^4 x \, e^{iqx} < 0 | T(A_\mu(x) A_\nu^\dagger(0)) | 0 >$$
$$= -g_{\mu\nu} \, \Pi_1(q^2) - q_\mu q_\nu \, \Pi_A(q^2) , \qquad (7.4)$$

with $V_\mu(x) =: \bar{d}(x) \gamma_\mu u(x)$: is the conserved vector current in the chiral limit, $A_\mu(x) =: \bar{d}(x) \gamma_\mu \gamma_5 u(x)$: the axial-vector current, and q_μ is the four-momentum carried by the currents. The functions $\Pi_{V,A}(q^2)$ are free of kinematical singularities, thus satisfying dispersion relations. In perturbative QCD (PQCD) they are normalized as

© The Author(s), under exclusive licence to Springer Nature Switzerland AG 2018
C. A. Dominguez, *Quantum Chromodynamics Sum Rules*,
SpringerBriefs in Physics, https://doi.org/10.1007/978-3-319-97722-5_7

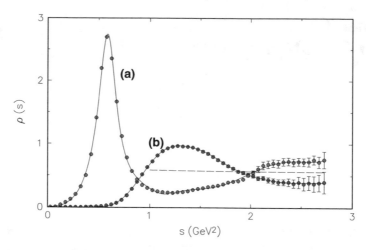

Fig. 7.1 ALEPH data [3] on the **a** vector, and **b** axial-vector spectral functions, $\rho(s) \equiv \mathrm{Im}\, \Pi_{V,A}(s)$, from τ-decay. The dash line is the PQCD prediction

$$\mathrm{Im}\, \Pi_V(q^2) = \mathrm{Im}\, \Pi_A(q^2) = \frac{1}{4\pi}\left[1 + \mathcal{O}\left(\alpha_s(q^2)\right)\right] . \tag{7.5}$$

A word of warning regarding Eq. (7.5), as it is common in the literature to normalize to an overall factor of $1/8\pi$. The difference stems from whether one employs the electric charge neutral current instead of the electrically charged current, as in Eqs. (7.3), (7.4).

In QCD, and in the chiral limit, $W_{1,2}$ become finite energy sum rules for $s > s_0 \simeq 1 - 3\,\mathrm{GeV}^2$, due to Eq. (7.5), plus the fact that at these energies the leading vacuum condensate contribution, i.e. the dimension $d = 6$ four-quark condensate, is negligible (the gluon condensate is chiral-symmetric). The WSR can then be written as

$$W_n(s_0) \equiv \int_0^{s_0} ds\, s^n \frac{1}{\pi}\left[\mathrm{Im}\Pi_V(s) - \mathrm{Im}\Pi_A(s)\right] = 2\, f_\pi^2\, \delta_{n0} \quad (n = 0, 1)\,, \tag{7.6}$$

where $s_0 \simeq 1 - 3\,\mathrm{GeV}^2$ is the squared energy beyond which QCD is valid. The WSR, Eq. (7.6), also follows from Cauchy's theorem in the complex s-plane, and the assumption of quark-hadron duality, i.e.

$$\int_0^{s_0} ds\, f(s)\, \frac{1}{\pi}\, \mathrm{Im}\Pi(s) = -\frac{1}{2\pi i}\oint_{|s|=s_0} f(s)\, \Pi(s)\, ds\,, \tag{7.7}$$

where $f(s) = s^n$ to reproduce Eq. (7.6).

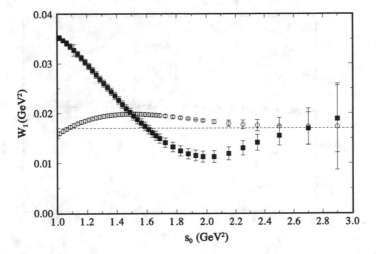

Fig. 7.2 The first WSR, Eq. (7.1), (solid squares), together with the modified (pinched) sum rule (open circles), Eq. (7.8), as a function of s_0, with the integrals computed from the ALEPH data [3]. The straight dotted line is the right-hand-side of Eq. (7.8), i.e. $2 f_\pi^2$

The ALEPH data on the vector and axial-vector spectral functions, determined from hadronic decays of the τ-lepton, are shown in Fig. 7.1. The resonance peaks correspond to the ρ-meson in the vector channel, and to the a_1 meson in the axial-vector channel. Agreement with QCD seems to be beyond the end-point of the data, although quark-hadron duality violation (DV) cannot be ruled out at this stage. However, in view of the results discussed in Chap. 5, on the extension of the kinematic range, one expects PQCD to be valid for $s \gtrsim 3 \, \text{GeV}^2$. This situation led to the proposal [4, 5] of introducing pinched integration kernels in the FESR, Eq. (7.7), in order to quench potential DV. The first WSR, Eq. (7.1), is replaced by a linear combination of the first and the second WSR, Eqs. (7.1) and (7.2)

$$\int_0^\infty ds \left(1 - \frac{s}{s_0}\right) \frac{1}{\pi} \left[\text{Im}\Pi_V(s) - \text{Im}\Pi_A(s)\right] = 2 f_\pi^2, \qquad (7.8)$$

with the normalization as in Eq. (7.5). Using ALEPH data on the spectral functions from τ-decay [3] in Eq. (7.8) gives the result shown in Fig. 7.2 (open circles), compared with the original sum rule Eq. (7.1) (solid squares). The impact of the pinched kernel is quite clear. It should be emphasized, though, that the lack of saturation in Eq. (7.1) does not necessarily imply DV. It could simply be due to the fact that the threshold for PQCD lies above the end-point of the τ-data.

Another chiral sum rule, the Das-Mathur-Okubo (DMO) sum rule [6], is obtained from the difference between the imaginary parts of the vector and the axial-vector correlator in the up- and down-quark sector, weighted by the kernel $1/s$, i.e.

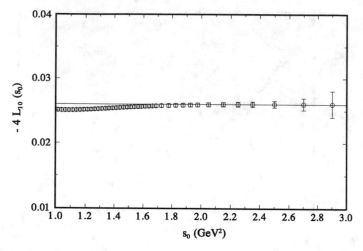

Fig. 7.3 $-4\,L_{10} \equiv \bar{\Pi}(0)$ as a function of s_0 from the pinched DMO sum rule, Eq. (7.11)

$$\bar{\Pi}(0) = \frac{1}{\pi} \int_0^{s_0} \frac{ds}{s} \left[\mathrm{Im}\, \Pi_V(s) - \mathrm{Im}\Pi_A(s) \right], \qquad (7.9)$$

with the pion-pole excluded from $\Pi_A(s)$. The finite remainder $\bar{\Pi}(0)$ is related to a counter-term of the order $\mathcal{O}(p^4)$ Lagrangian of CHPT [7], $\bar{L}_{10}$, and can be expressed as

$$\bar{\Pi}(0) = -4\,\bar{L}_{10} = \frac{1}{3} \left[f_\pi^2 \langle r_\pi^2 \rangle - \frac{1}{2} F_A \right] = 0.026 \pm 0.001, \qquad (7.10)$$

where $\langle r_\pi^2 \rangle = 0.439 \pm 0.008\,\mathrm{fm}^2$ is the electromagnetic radius of the pion [8], and $F_A = 0.0119 \pm 0.0001$ is the radiative pion decay constant [9].

The DMO sum rule, Eq. (7.9), is reasonably saturated by the ALEPH data above $s_0 \simeq 1.5\,\mathrm{GeV}^2$. A *pinched* version of this sum rule, proposed in [5, 10], is

$$\bar{\Pi}(0) = 2\,\frac{f_\pi^2}{s_0} + \int_0^{s_0} \frac{ds}{s} \left(1 - \frac{s}{s_0}\right)^2 \frac{1}{\pi} \left[\mathrm{Im}\, \Pi_V(s) - \mathrm{Im}\Pi_A(s) \right]. \qquad (7.11)$$

This pinched sum rule is also well satisfied above $s_0 \simeq 1.5\,\mathrm{GeV}^2$, as shown in Fig. (7.3) giving [5]

$$\bar{L}_{10} = -(6.5 \pm 0.1) \times 10^{-3}, \qquad (7.12)$$

a result in very good agreement with an earlier determination [10], $\bar{L}_{10} = -(6.43 \pm 0.08) \times 10^{-3}$, as well as with other determinations from more involved procedures to deal with potential DV [11, 12], $\bar{L}_{10} = -(6.46 \pm 0.15) \times 10^{-3}$. It also agrees with lattice QCD determinations, albeit within their larger uncertainties [13, 14].

References

1. V. de Alfaro, S. Fubini, G. Furlan, C. Rossetti, *Currents in Hadron Physics* (Noth Holland, Amsterdam, 1973)
2. S. Weinberg, Phys. Rev. Lett. **18**, 507 (1967)
3. M. Davier, A. Höcker, Z. Zhang, Rev. Mod. Phys. **78**, 1043 (2006)
4. K. Maltman, Phys. Lett. B **440**, 367 (1998)
5. C.A. Dominguez, K. Schilcher, Phys. Lett. B **581**, 193 (2004)
6. T. Das, V.S. Mathur, S. Okubo, Phys. Rev. Lett. 19, 859 (1967); T. Das, G.S. Guralnik, V.S. Mathur, F.E. Low, J.E. Young. Phys. Rev. Lett. **18**, 759 (1967)
7. S. Scherer, Adv. Nucl. Phys. **27**, 277 (2003); J. Gasser, H. Leutwyler, Nucl. Phys. B **250**, 465 (1985); G. Ecker, J. Gasser, A. Pich, E. de Rafael, Nucl. Physi. B **321**, 311 (1989); G. Amoros, J. Bijnens, P. Talavera, Nucl. Phys. B **585**, 293 (2000) [Erratum *ibid.* **598**, 665 (2001)]
8. S.R. Amendolia et al., Nucl. Phys. B **277**, 168 (1986)
9. K.G. Patrignani et al., Particle data group. Chinese Phys. C **40**, 100001 (2016)
10. C.A. Dominguez, K. Schilcher, Phys. Lett. B **448**, 93 (1999)
11. M. Gonzalez-Alonso, A. Pich, J. Prades, Phys. Rev. D **81**, 074007 (2010)
12. M. Gonzalez-Alonso, A. Pich, J. Prades, Phys. Rev. D **82**, 014019 (2010)
13. JLQCD collaboration, E. Shintani et al., Phys. Rev. Lett. **101**, 242001 (2008)
14. RBC, UKQCD collaboration, P.A. Boyle, L. Del Debbio, J. Wennekers, J.M. Zanotti, Phys. Rev. D **81**, 014504 (2010)

Chapter 8
Determination of the Gluon Condensate

As mentioned in Chap. 2, the gluon condensate of dimension $d = 4$ is one of the two leading order terms in the OPE, Eq. (2.2), together with the light-quark condensate. The gluon condensate was introduced in the pioneer papers on Laplace QCD sum rules by Shifman et al. [1]. Its numerical value was estimated using experimental information on $e^- e^+$ annihilation in the charmonium channel, with the result [1]

$$\left\langle \frac{\alpha_s}{\pi} G^2 \right\rangle = 0.012 \, \text{GeV}^4 \,, \tag{8.1}$$

with no uncertainty given. This value remained in use in applications for quite some time. A detailed analysis, also in the charmonium channel, with determined uncertainties, was performed in [2] with the result

$$\left\langle \frac{\alpha_s}{\pi} G^2 \right\rangle = 0.014 \pm 0.0044 \, \text{GeV}^4 \,. \tag{8.2}$$

This result was questioned in [3, 4] claiming it underestimates its value by up to a factor three. British groups [5, 6] confirmed this claim in the framework of two-dimensional QCD, and a reanalysis in the charmonium channel [7, 8] reached similar conclusions. Using experimental data on electron-positron annihilation into hadrons, and Laplace transform QCD sum rules, the gluon condensate was found to be one order of magnitude smaller than Eq. (8.1) [9]. Also using experimental data on electron-positron annihilation into hadrons, within FESR [10], a bound was obtained

$$0.07 \, \text{GeV}^4 \leq \left\langle \frac{\alpha_s}{\pi} G^2 \right\rangle \leq 0.19 \, \text{GeV}^4 \,. \tag{8.3}$$

A determination using the first set of ALEPH data on the vector and axial-vector current correlators from hadronic decays of the τ-lepton [11], together with FESR, found [12]

© The Author(s), under exclusive licence to Springer Nature Switzerland AG 2018
C. A. Dominguez, *Quantum Chromodynamics Sum Rules*,
SpringerBriefs in Physics, https://doi.org/10.1007/978-3-319-97722-5_8

$$\left\langle \frac{\alpha_s}{\pi} G^2 \right\rangle = (0.1 - 0.4)\,\text{GeV}^4 . \tag{8.4}$$

This historical account is provided solely for the purpose of highlighting the difficulties encountered in determining the value of the gluon condensate.

To add to the confusion, and on the purely theoretical domain, there have been claims that the gluon condensate is affected by *renormalon ambiguities*, and thus its numerical value remains in limbo [13]. This is definitely not the case, for a variety of reasons. First, and foremost, the gluon condensate is a renormalization group invariant quantity in the operator product expansion of current correlators at short distances, Eq. (2.2), to be interpreted solely as a parameter of the QCD sum rule method [14]. Its numerical value is to be obtained only by confronting QCD sum rules with experimental data. Hence, no further claims are made on its nature/origin. Second, the renormalon issue is severely model-dependent, in that its value relies on estimates of extremely high order contributions in perturbative QCD (similar to ladder diagrams). Clearly, estimates are not determinations.

Regarding the numerical value of the gluon condensate, all the estimates and determinations mentioned above are rather dated, as the QCD information at the time of their determination was restricted to next or to next-to-next to leading order in perturbative QCD, i.e. to $\mathcal{O}(\alpha_s)$ or $\mathcal{O}(\alpha_s^2)$. Current information is up to $\mathcal{O}(\alpha_s^4)$ in the light-quark region, and $\mathcal{O}(\alpha_s^3)$ in the charm- and bottom-quark regions. Furthermore, the procedures to determine the gluon condensate required the strong coupling, α_s, dependent on the value of Λ_{QCD} entering the logarithm, i.e.

$$\alpha_s(s_0) = \frac{2\pi}{b_0 \ln\left(s_0/\Lambda_{QCD}^2\right)} + \cdots , \tag{8.5}$$

where b_0 is given in Eq. (3.3). The value of Λ_{QCD} at the time of the above determinations was in the range $\Lambda_{QCD} \simeq 100 - 150$ MeV, while currently it is more than twice as large! [15].

Turning to the present, the most recent QCD-FESR determinations of the gluon condensate are (i) based on data on $e^+ e^-$ annihilation to hadrons in the region $(1.0 - 5.0)\,\text{GeV}^2$ using standard FESR [16], and (ii) based on latest ALEPH data on hadronic spectral functions [17], and (iii) from a novel, non-standard FESR in the charm-quark region, designed to overcome the drawbacks of standard FESR, and allowing for unprecedented and trustworthy accuracy [18]. The latter also allows to test the expectation that the gluon condensate should be scale-independent.

A description of the most recent gluon condensate determination from data on $e^+ - e^-$ annihilation into hadrons [16] starts with the following definitions. The relevant two-point function in this case is the electromagnetic current correlator

$$\Pi_{\mu\nu}^{\text{EM}}(q^2) = i \int d^4x\, e^{iqx}\, \langle 0|T(J_{\mu}^{\text{EM}}(x)\, J_{\nu}^{\text{EM}}(0))|0\rangle$$
$$= \left(-q^2 g_{\mu\nu} + q_\mu q_\nu\right) \Pi^{\text{EM}}(q^2) , \tag{8.6}$$

where for three flavours

$$J_\mu^{EM}(x) = \frac{2}{3}\bar{u}(x)\gamma^\mu u(x) - \frac{1}{3}\bar{d}(x)\gamma^\mu d(x) - \frac{1}{3}\bar{s}(x)\gamma^\mu s(x) \, . \tag{8.7}$$

Since QCD is flavour blind, and if isospin invariance is exact, then it is convenient to define a QCD current correlator using any of the quark currents $\bar{q}_i\gamma_\mu q_i$ with flavour i. This leads to

$$\text{Im}\,\Pi^{EM}(q^2) = \sum_{i=1}^{n_f} Q_i^2 \,\text{Im}\,\Pi_{VV}(q^2) \, , \tag{8.8}$$

with Q_i the charge of the quark $i = u, d, s, \ldots$, $\sum_i Q_i^2 = 2/3$ for $n_f = 3$, and $\text{Im}\,\Pi_{VV}(q^2)$ is the QCD correlator of vector currents of flavour i. The spectral function, $\text{Im}\,\Pi^{EM}(s)$, with s the square energy, is accessible experimentally from data on e^+e^- annihilation into hadrons as follows.

Starting in the hadronic sector, the standard ratio $R(s)$ is

$$R(s) = \frac{\sigma_{TOT}(e^+e^- \to \text{hadrons})}{\sigma(e^+e^- \to \mu^+\mu^-)} \, , \tag{8.9}$$

where the electromagnetic cross section is

$$\sigma(e^+e^- \to \mu^+\mu^-) = \frac{4\pi\alpha_{EM}^2}{3s} \, , \tag{8.10}$$

and $\alpha_{EM} = e^2/4\pi$. In QCD the relation between R and the electromagnetic spectral function is given by

$$R(s) = 12\pi\,\text{Im}\,\Pi^{EM}(s) = 3\sum_{i=1}^{n_f} Q_i^2 \left(1 + \frac{\alpha_s}{\pi} + \ldots\right) . \tag{8.11}$$

A singlet contribution proportional to $\left(\sum_i Q_i\right)^2$ arises at order $\mathcal{O}(\alpha_s^3)$ and vanishes if one sums over three flavours. In the case of the two-pion final state, dominated by the ρ-resonance, there is a relation between R and the pion form factor, $F_\pi^{(0)}(s)$

$$R_{e^+e^- \to \pi^+\pi^-}(s) = \frac{1}{4}\left(1 - \frac{4m_\pi^2}{s}\right)^{\frac{3}{2}} |F_\pi^{(0)}(s)|^2 \, . \tag{8.12}$$

Turning to the QCD sector, the vector current correlator, $\Pi_{VV}(q^2)$, in Eq. (8.8) is given by the OPE, Eq. (2.2). Of particular interest here is the purely perturbative term, as well as the dimension $d = 4$ non-perturbative contribution. The latter involves the gluon condensate together with the quark condensate multiplied by the quark masses

$$C_4 \langle \mathcal{O}_4 \rangle = \frac{\pi^2}{3} \langle \frac{\alpha_s}{\pi} G_{\mu\nu} G^{\mu\nu} \rangle + 4\pi^2 \left(m_u \langle \bar{u}u \rangle + m_d \langle \bar{d}d \rangle + m_s \langle \bar{s}s \rangle \right), \qquad (8.13)$$

where α_s is the running strong coupling, and in the sequel $\langle 0|\mathcal{O}_{2N}|0 \rangle \equiv \langle \mathcal{O}_{2N} \rangle$ is to be understood. This condensate is renormalization group invariant to all orders in PQCD.

The purely perturbative QCD part of $\mathrm{Im}\,\Pi_{VV}(q^2)$ is currently known [19, 20] up to order $\mathcal{O}(\alpha_s^4)$

$$8\pi \, \mathrm{Im}\, \Pi_{VV}(s) = 1 + a_s + a_s^2 \left(F_3 + \frac{b_1}{2} L_\mu \right) + a_s^3 \left[F_4 + \left(b_1 F_3 + \frac{b_2}{2} \right) L_\mu \right.$$
$$\left. + \frac{b_1^2}{4} L_\mu^2 \right] + a_s^4 \left[k_3 - \frac{\pi^2}{4} b_1^2 F_3 - \frac{5}{24} \pi^2 b_1 b_2 + \left(\frac{3}{2} b_1 F_4 + b_2 F_3 + \frac{b_3}{2} \right) L_\mu \right.$$
$$\left. + \frac{b_1}{2} \left(\frac{3}{2} b_1 F_3 + \frac{5}{4} b_2 \right) L_\mu^2 + \frac{b_1^3}{8} L_\mu^3 \right], \qquad (8.14)$$

where $a_s \equiv \alpha_s(\mu^2)/\pi$, $L_\mu \equiv \ln(Q^2/\mu^2)$, $k_3 = 49.076$ [21], $F_3 = 1.9857 - 0.1153\, n_f$, and $F_4 = 18.2427 - \frac{\pi^2}{3}(\frac{b_1}{2})^2 - 4.2158\, n_f + 0.0862\, n_f^2$, with b_1 given in Eq. (3.4). Notice that the normalization factor 8π is due to the nature of the vector current, Eq. (8.7), entering in the correlator Eq. (8.6) (recall that for electrically charged currents the standard normalization factor is instead 4π). The expression for the running coupling up to five-loop order is given by [22]

$$\frac{\alpha_s^{(4)}(s_0)}{\pi} = \frac{\alpha_s^{(1)}(s_0)}{\pi} + \left(\frac{\alpha_s^{(1)}(s_0)}{\pi} \right)^2 \left(-\frac{b_2}{b_1} \ln L \right)$$
$$+ \left(\frac{\alpha_s^{(1)}(s_0)}{\pi} \right)^3 \left(\frac{b_2^2}{b_1^2} \left(\ln^2 L - \ln L - 1 \right) + \frac{b_3}{b_1} \right)$$
$$- \left(\frac{\alpha_s^{(1)}(s_0)}{\pi} \right)^4 \left[\frac{b_2^3}{b_1^3} \left(\ln^3 L - \frac{5}{2} \ln^2 L - 2 \ln L + \frac{1}{2} \right) \right.$$
$$\left. + 3 \frac{b_2 b_3}{\beta_1^2} \ln L + \frac{b_3}{b_1} \right], \qquad (8.15)$$

with

$$\frac{\alpha_s^{(1)}(s_0)}{\pi} \equiv \frac{-2}{\beta_1 L}, \qquad (8.16)$$

where $L \equiv \ln \left(s_0 / \Lambda_{\overline{\mathrm{MS}}}^2 \right)$ defines the standard $\overline{\mathrm{MS}}$ scale $\Lambda_{\overline{\mathrm{MS}}} = (340 \pm 8)\,\mathrm{MeV}$ for $n_F = 3$ [15].

The FESR in this channel can be written as

$$(-)^N C_{2N+2} \langle \mathcal{O}_{2N+2} \rangle = 8\pi^2 \int_0^{s_0} ds \, s^N \frac{1}{\pi} \mathrm{Im}\, \Pi^{\mathrm{DATA}}(s) - s_0^{N+1} M_{2N+2}(s_0), \qquad (8.17)$$

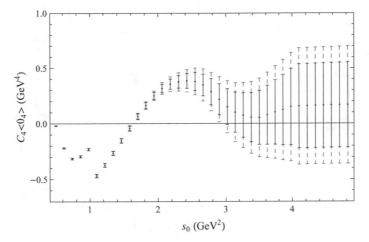

Fig. 8.1 $C_4 \langle \mathcal{O}_4 \rangle$ calculated in FOPT using $\alpha_s(M_\tau^2) = 0.321 \pm 0.015$, corresponding to $\Lambda_{\overline{\text{MS}}}^{(n_f=3)} = 341$ MeV. The smaller uncertainties are obtained assuming no correlations between experiments, while the larger ones assume 100% correlations for data obtained using the same experimental facility

where the dimensionless PQCD moments $M_{2N+2}(s_0)$ are given by

$$
\begin{aligned}
M_{2N+2}(s_0) &\equiv -8\pi^2 \frac{1}{2\pi i} \oint_{C(|s_0|)} \frac{ds}{s_0} \left[\frac{s}{s_0} \right]^N \Pi_{VV}(s) \\
&= 8\pi^2 \int_0^{s_0} \frac{ds}{s_0} \left[\frac{s}{s_0} \right]^N \frac{1}{\pi} \operatorname{Im} \Pi_{VV}(s) .
\end{aligned} \tag{8.18}
$$

Given the expected size of the dimension-four term, $C_4\langle \mathcal{O}_4 \rangle$, Eq. (8.13), i.e. $C_4\langle \mathcal{O}_4 \rangle = \mathcal{O}(10^{-3}\,\text{GeV}^4)$, one expects a sizeable cancellation between the data contribution and PQCD in Eq. (8.17). A posteriori, this turns out to be the source of the large uncertainty in the result, as shown in Fig. 8.1 using FOPT. The hadronic data used in the line integral in Eq. (8.17) is described in detail in [16]. It includes twenty-nine different sources in a variety of hadronic final-states.

The situation with the dimension $d = 6$ condensate is even worse, as expected. The results for $C_6 \langle \mathcal{O}_6 \rangle$ are shown in Fig. 8.2 in a different vertical scale. The origin and structure of $C_6 \langle \mathcal{O}_6 \rangle$ is discussed at the end of this section.

The conclusion is that this approach is unsuccessful. Even if more accurate data were available, the partial cancellation between the two terms of comparable magnitude on the right hand side of Eq. (8.17) is a serious challenge.

Another attempt at obtaining the value of the gluon condensate was made in [17], still from QCD FESR but using instead the ALEPH data base from τ-decay [23]. The relevant correlators are

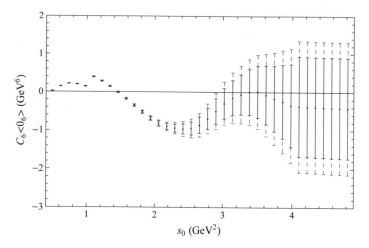

Fig. 8.2 $C_6 \langle \mathcal{O}_6 \rangle$ calculated in FOPT using $\alpha_s(M_\tau^2) = 0.321 \pm 0.015$, corresponding to $\Lambda_{\overline{\text{MS}}}^{(n_f=3)} = 341$ MeV. The smaller uncertainties are obtained assuming no correlations between experiments, while the larger ones assume 100% correlations for data obtained using the same experimental facility

$$\Pi_{\mu\nu}^{VV}(q^2) = i \int d^4x \, e^{iqx} \langle 0|T(V_\mu(x)V_\nu^\dagger(0))|0\rangle \tag{8.19}$$
$$= (-g_{\mu\nu} \, q^2 + q_\mu q_\nu) \, \Pi_V(q^2) \,,$$

$$\Pi_{\mu\nu}^{AA}(q^2) = i \int d^4x \, e^{iqx} \langle 0|T(A_\mu(x)A_\nu^\dagger(0))|0\rangle \tag{8.20}$$
$$= (-g_{\mu\nu}q^2 + q_\mu q_\nu) \, \Pi_A(q^2) - q_\mu q_\nu \, \Pi_0(q^2) \,,$$

where $V_\mu(x) =: \bar{u}(x)\gamma_\mu d(x) :$, $A_\mu(x) =: \bar{u}(x)\gamma_\mu\gamma_5 d(x) :$, with $u(x)$ and $d(x)$ the quark fields, and $\Pi_{V,A}(q^2)$ normalized in perturbative QCD (PQCD) (in the chiral limit) according to

$$\frac{1}{\pi} \text{Im} \, \Pi_V^{PQCD}(s) = \frac{1}{\pi} \text{Im} \, \Pi_A^{PQCD}(s) = \frac{1}{4\pi^2}\left(1 + \frac{\alpha_s(s)}{\pi} + ...\right) , \tag{8.21}$$

where $s \equiv q^2 > 0$ is the squared energy. Notice the normalization factor due to the currents being electrically charged. The Lorentz decomposition is used to separate the correlation function into its $J = 1$ and $J = 0$ parts. To the accuracy needed in the following, the vector current can be assumed to be conserved. The running strong coupling was obtained from Eq. (3.9) using as input [24] $\alpha_s(M_\tau^2) = 0.341 \pm 0.013$.

Instead of using the correlators Eqs. (8.19) and (8.20), the Adler function was invoked as it is more convenient when using contour improved perturbation theory (CIPT) to integrate in the complex plane. The Adler function is defined as

$$D(s) \equiv -s \frac{d}{ds} \Pi(s) , \tag{8.22}$$

with $\Pi(s) \equiv \Pi_{V,A}(s)$. Invoking Cauchy's theorem and after integration by parts the following relation is obtained

$$\oint_{|s|=s_0} ds \left(\frac{s}{s_0}\right)^N \Pi(s) = \frac{1}{N+1} \frac{1}{s_0^N} \oint_{|s|=s_0} \frac{ds}{s} \left(s^{N+1} - s_0^{N+1}\right) D(s) . \tag{8.23}$$

After RG improvement, the perturbative expansion of the Adler function becomes

$$D(s)^{\cdot} = \frac{1}{4 \pi^2} \sum_{m=0} K_m \left[\frac{\alpha_s(-s)}{\pi}\right]^m , \tag{8.24}$$

where $K_0 = K_1 = 1$, $K_2 = 1.6398$, $K_3 = 6.3710$, for three flavours, and $K_4 = 49.076$ [19–22]. The vacuum condensates are determined from the following pinched FESR

$$C_{2N+2} \langle O_{2N+2} \rangle = (-)^{N+1} 4\pi^2 s_0^N \int_0^{s_0} ds \left[1 - \left(\frac{s}{s_0}\right)^N\right] \frac{1}{\pi} \mathrm{Im}\ \Pi(s)^{HAD}$$

$$+ (-)^N s_0^{N+1} \left[M_0(s_0) - M_N(s_0)\right] , \tag{8.25}$$

$$M_N(s_0) = \frac{1}{2\pi} \frac{1}{(N+1)} \sum_{m=0} K_m \left[I_{N+1,m}(s_0) - I_{0,m}(s_0)\right] , \tag{8.26}$$

with

$$I_{N,m} \equiv i \oint_{|s|=s_0} ds \left(\frac{s}{s_0}\right)^N \left[\frac{\alpha_s(-s)}{\pi}\right]^m , \tag{8.27}$$

to be computed as explained in Chap. 4.

A key issue with the FESR Eq. (8.25) is the opposite signs of the two terms on its right-hand-side. In practice these two terms are roughly of the same order of magnitude, thus leading to results affected by large uncertainties.

The results for $C_4 \langle O_4 \rangle$ in CIPT are shown in Fig. 8.3 for $\alpha_s(M_\tau^2) = 0.341$ from [24]. The data near the end-point ($s_0 \simeq M_\tau^2$) are affected by such uncertainties that no safe estimate of the gluon condensate can be given there. However, there is a region $s_0 \simeq 2.0 - 2.4 \,\mathrm{GeV}^2$ where an estimate can be obtained, albeit with a large error, i.e.

$$C_4 \langle O_4 \rangle |_{\mathrm{CIPT}} = (0.017 \pm 0.012) \,\mathrm{GeV}^4 , \tag{8.28}$$

where this value is obtained from results corresponding to the $V + A$ spectral function at $s_0 = 2.35 \,\mathrm{GeV}^2$. Results from using FOPT are

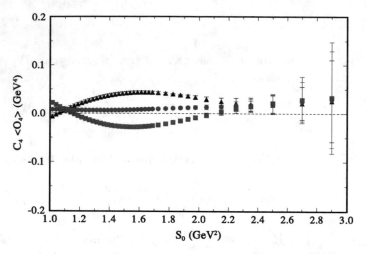

Fig. 8.3 The dimension $d = 4$ condensate in CIPT from the FESR, Eq. (8.25), with $N = 1$ and $\alpha_s(M_\tau^2) = 0.341$ from [24]. The ALEPH data for the vector (V) (upper black solid triangles), the axial (A) (lower blue solid squares) and the average $\frac{1}{2}(V + A)$ (middle red solid dots) spectral function were used

$$C_4\langle O_4\rangle|_{\text{FOPT}} = (0.022 \pm 0.006)\,\text{GeV}^4 \,. \tag{8.29}$$

Combining both results leads to the conservative upper bound [17]

$$C_4\langle O_4\rangle|_{V,A} \lesssim 0.035\,\text{GeV}^4 \,. \tag{8.30}$$

It should be mentioned that a very different result was obtained by the ALEPH collaboration [23] after fitting simultaneously all relevant parameters, obtaining negative values for the gluon condensate from the axial-vector channel, as well as from the $V + A$ channel. Given the positivity of $C_4\langle O_4\rangle$, this global fitting procedure seems unreliable.

An entirely different and successful approach to determine the gluon condensate, still based on QCD FESR, was proposed in [18]. This leads to a reasonably accurate value in full agreement with the bound Eq. (8.30). The first consideration is to choose the charm-quark energy region, where data for the R-ratio in $e^+e^- \to hadrons$ [25, 26] starts with the two well known narrow-width resonances J/ψ and $\psi(2S)$. Next, the approach should be such that this contribution to a FESR becomes leading. The latter excludes the standard FESR, e.g. Eq. (8.17), as it lacks such weighting. The optimal FESR satisfying this criterion should be such that the gluon condensate does not become the result of a cancellation between terms of similar magnitude, as it happens in Eq. (8.17). In order to achieve this one starts by considering the most general FESR

$$\int_0^{s_0} p(s) \frac{1}{\pi} Im \, \Pi(s) \, ds = -\frac{1}{2\pi i} \oint_{C(|s_0|)} p(s) \, \Pi(s) \, ds + \text{Res}[\Pi(s) \, p(s), s = 0] \,,$$

$$(8.31)$$

where $p(s)$ is now a meromorphic function, the integral on the right hand side involves QCD, provided s_0 is large enough, and the left hand side involves the hadronic spectral function

$$Im \, \Pi(s)|_{DATA} = \frac{1}{12\pi} R_c(s) \,, \qquad (8.32)$$

with $R_c(s)$ the standard R-ratio for charm production in e^+e^- annihilation. The residue term in Eq. (8.31) would appear if $p(s)$ is singular at some point(s) in the complex plane, e.g. if $p(s)$ is of the form

$$p(s) = \frac{1}{s^{N+1}} \,. \qquad (8.33)$$

The leading non-perturbative contribution to a FESR involving such a kernel was obtained in [27], and in [28] using the $\bar{MS}$-scheme with the result

$$\text{Res}\left[\frac{\Pi(s)|_{NPQCD}}{s^{N+1}}, s = 0\right] = \frac{e_c^2}{(4\bar{m}_c^2)^{N+2}} \left\langle \frac{\alpha_s}{\pi} G^2 \right\rangle a_N \left(1 + \frac{\alpha_s}{\pi} \bar{b}_N\right), \qquad (8.34)$$

where the charm-quark mass, $\bar{m}_c$, and the strong coupling α_s depend on the renormalization scale μ, and

$$a_N = -\frac{2N+2}{15} \frac{\Gamma(4+N)\,\Gamma(7/2)}{\Gamma(7/2+N)\,\Gamma(4)}, \qquad (8.35)$$

$$\bar{b}_N = b_N - (2N+4)\left(\frac{4}{3} - l_m\right), \qquad (8.36)$$

where $l_m \equiv \ln(\bar{m}_c^2(\mu)/\mu^2)$, μ is a renormalization scale, $b_0 = 1469/162$, $b_1 = 135779/12960$, and $b_2 = 1969/168$. Other values of b_i are given in [27, 28]. The unknown NNLO term will be included as a source of uncertainty in the final analysis. The QCD parameters are the charm-quark mass $m_c(\mu^2)$, the strong coupling $\alpha_s(\mu^2)$, and the gluon condensate $\langle \frac{\alpha_s}{\pi} G^2 \rangle$. Their numerical values used in [18] are: $\alpha_s(M_Z^2) = 0.1183 \pm 0.0007$ from LQCD [29], to be scaled down to the charm-quark region using Eq. (3.9), and the charm-quark mass from LQCD [30] $\bar{m}_c(3\,\text{GeV}) = 986.4 \pm 4.1\,\text{GeV}$, in good agreement with the most recent QCDSR determination [31] $\bar{m}_c(3\,\text{GeV}) = 987 \pm 9\,\text{MeV}$.

There is also a low-energy PQCD contribution to be taken into account, together with the gluon condensate. The QCD low energy expansion in inverse powers of the charm-quark mass can be formally written as

$$\Pi_{PQCD}(s) = \frac{3\,e_c^2}{16\,\pi^2} \sum_{n\geq 0} \bar{C}_n\, z^n ,$$ (8.37)

where $z = s/(4\bar{m}_c^2)$. The coefficients $\bar{C}_n$ are then expanded in powers of $\alpha_s(\mu)$

$$\bar{C}_n = \bar{C}_n^{(0)} + \frac{\alpha_s(\mu)}{\pi} \left(\bar{C}_n^{(10)} + \bar{C}_n^{(11)} l_m\right) + \left(\frac{\alpha_s(\mu)}{\pi}\right)^2 \left(\bar{C}_n^{(20)} + \bar{C}_n^{(21)} l_m + \bar{C}_n^{(22)} l_m^2\right)$$
$$+ \left(\frac{\alpha_s(\mu)}{\pi}\right)^3 \left(\bar{C}_n^{(30)} + \bar{C}_n^{(31)} l_m + \bar{C}_n^{(32)} l_m^2 + \bar{C}_n^{(33)} l_m^3\right) + \cdots .$$ (8.38)

where $l_m \equiv \ln(\bar{m}_c^2(\mu)/\mu^2)$. Up to three loop level the coefficients of $\bar{C}_n$ are known up to $n = 30$ [32, 33]. At four-loop level $\bar{C}_0$ and $\bar{C}_1$ were determined in [32, 34], [35], $\bar{C}_2$ is from [33, 36], and $\bar{C}_3$ from [37]. The kernel $p(s)$ is chosen so that no coefficients $\bar{C}_4$ and above contribute to the Cauchy residue at $s = 0$.

Notice that the gluon condensate also enters in the contour integral, Eq. (8.31). However, its contribution to that integral is negligible. Its appearance in the residue, Eq. (8.34), is the key point of this approach. Indeed, substituting Eq. (8.34) into the FESR Eq. (8.31) gives

$$\left\langle \frac{\alpha_s}{\pi} G^2 \right\rangle = \frac{(4\bar{m}_c^2)^{N+2}}{e_c^2\, a_N \left(1 + \frac{\alpha_s}{\pi} \bar{b}_N\right)} \left[\int_0^{s_0} p(s)\, \frac{1}{\pi} Im\, \Pi(s)\, ds \right.$$
$$\left. + \frac{1}{2\pi i} \oint_{C(|s_0|)} p(s)\Pi(s) ds - Res\left(\Pi_{PQCD}(s)\, p(s)\right)\Big|_{s=0} \right],$$ (8.39)

where the last term above is the residue at the singularity from the low energy PQCD expansion, Eq. (8.37). It is non-zero depending on the value of N in the integration kernel $p(s)$, Eq. (8.33). A key property of this FESR is that the two integral terms in brackets, i.e. the data line integral, and the contour QCD integral now have the same sign !

Solving the renormalization group equation for the strong coupling and for the quark mass one can obtain their values at any scale s in terms of their values at any given reference scale s^* from Eqs. (3.9) and (3.31), e.g. $s = s_0$. Regarding the renormalization scale μ, we follow the choice [31–38] $\mu^2 = (3\,\text{GeV})^2$ in the low energy QCD expansion, and $\mu^2 = s_0$ in the high energy QCD expansion on the circle of radius $s = |s_0|$.

The PQCD correlator entering the contour integral in Eq. (8.39) is given by the (high energy) expansion

$$\Pi(s)|_{PQCD} = e_c^2 \sum_{n=0}^{\infty} \left(\frac{\alpha_s(\mu^2)}{\pi}\right)^n \Pi^{(n)}(s) ,$$ (8.40)

where $e_c = 2/3$ is the charm-quark electric charge, and

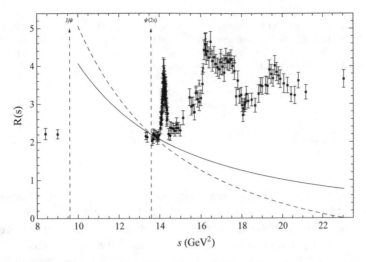

Fig. 8.4 Experimental data for $R(s)$ [25, 26] together with the optimal integration kernel, Eq. (8.42), with $N = 2$ (dash curve), and $p(s) = 1/s^2$ (solid curve) normalized to coincide with the former at the position of the $\psi(2S)$ peak

$$\Pi^{(n)}(s) = \sum_{i=0} \left(\frac{\bar{m}_c^2}{s}\right)^i \Pi_i^{(n)} , \qquad (8.41)$$

with $\bar{m}_c \equiv \bar{m}_c(\mu)$ the running charm-quark mass in the $\overline{MS}$-scheme. The function $\Pi(s)_{PQCD}$ is known up to next-to-next to leading order in PQCD [21, 39, 40].

Finally, turning to the experimental data in this region, shown in Fig. 8.4, the analysis followed that of [28, 41], to wit.

For the contribution of the first two narrow resonances PDG data [15] was used, followed by the open charm region, after subtraction of the light-quark sector contribution [42]. In the region 3.97 GeV $\leq \sqrt{s} \leq$ 4.26 GeV only CLEO data [25] was taken into account, as it is the more accurate.

Next, there are two data sets from BES [26, 43], which were assumed not fully independent for the error analysis, thus adding errors linearly, rather than in quadrature. These data set, though, is independent from CLEO [25] so that errors were combined in quadrature. There is a data *desert* region for $s = 25 - 49$ GeV^2, followed by CLEO data up to $s \simeq 90$ GeV^2, fully compatible with PQCD. For further details see [18].

Turning to the integration kernel, $p(s)$, Eq. (8.33), the optimal choice is

$$p(s) = \left(\frac{s_0}{s}\right)^N - 1 , \qquad (8.42)$$

with $N \geq 2$. A detailed justification of this choice is given in [18].

Table 8.1 Results for the gluon condensate for the kernel, Eq. (8.42), for $N = 2$ and its sources of uncertainty from s_0, α_s, m_c, the experimental data, and the total uncertainty. Method (a) refers to using the currently known NLO radiative correction to the residue, Eq. (8.34). Method (b) assumes that the NNLO correction is as large, and of the same sign as the NLO one

Method	$\langle \frac{\alpha_s}{\pi} G^2 \rangle$ (GeV^4)	Uncertainties (GeV^4)				
		Δ_{s_0}	Δ_{α_s}	Δ_{m_c}	Δ_{DATA}	Δ_{T}
(a)	0.044	0.0028	0.0003	0.0048	0.0043	0.007
(b)	0.026	0.0016	0.0001	0.0027	0.0024	0.004

The impact of the integration kernel, Eq. (8.42), can be appreciated from Fig. 8.4. As expected by design, the contribution of the data region to the line integral in the FESR, Eq. (8.39), enhances the well known first two narrow resonances, J/ψ and $\psi(2S)$, and quenches substantially the rest. The results for the gluon condensate together with the various uncertainties, is shown in Table 8.1. The result for the gluon condensate, after considering all uncertainties, is

$$\left\langle \frac{\alpha_s}{\pi} G^2 \right\rangle = 0.037 \pm 0.015 \, \text{GeV}^4 \,, \tag{8.43}$$

which agrees with a LQCD determination [44] $\langle \frac{\alpha_s}{\pi} G^2 \rangle = 0.028 \pm 0.003 \, \text{GeV}^4$. It is also consistent with the bound, Eq. (8.30). An independent QCD sum rule determination in the light-quark region, from an unconventional method, gives [45]

$$\langle \frac{\alpha_s}{\pi} G^2 \rangle = 0.062 \pm 0.019 \, \text{GeV}^4 \,, \tag{8.44}$$

in agreement with Eq. (8.43) within very large errors. If taken with caution, this result from the low energy region together with Eq. (8.43) from the high energy domain would support the scale independence of the gluon condensate.

Returning to the issue of the dimension $d = 6$ condensates, there are three possible structures built from the QCD quark and gluon fields

$$\hat{\mathcal{O}}_6|_q = \bar{q}(x) \, \Gamma_1 \, q(x) \, \bar{q}(x) \, \Gamma_2 \, q(x) \,, \tag{8.45}$$

$$\hat{\mathcal{O}}_6|_\sigma = m \, \bar{q}(x) \, \sigma_{\mu\nu} \frac{\lambda^a}{2} q(x) \, G^a_{\mu\nu}(x) \,, \tag{8.46}$$

$$\hat{\mathcal{O}}_6|_{G^3} = f_{abc} \, G^a_{\mu\nu}(x) \, G^b_{\nu\alpha}(x) \, G^c_{\alpha\mu}(x) \,, \tag{8.47}$$

corresponding to the four-quark condensate, the mixed quark-gluon condensate, and the three-gluon condensate. The object Γ_i in Eq. (8.45) above is a combination of gamma-matrices, SU(3) λ-matrices, and the tensor $\sigma_{\mu\nu} \propto [\gamma_\mu , \gamma_\nu]$, depending on the correlator (see [1]). None of these condensates shares the status of the quark condensate, which enters in the Gell-Mann-Oakes-Renner relation. In principle they

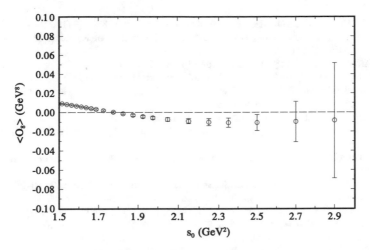

Fig. 8.5 The chiral condensate of dimension $d = 8$ from a pinched FESR [17]

would all contribute to the dimension $d = 6$ term in the OPE, Eq. (2.2), with a-priori unknown hierarchy. In the early days of the QCD sum rule technique attempts were made to relate the four-quark condensate to the square of the quark condensate through some sort of vacuum saturation ansatz [1]. This was accompanied by hopes that the other two $d = 6$ condensates would be numerically negligible. The vacuum saturation ansatz has been recently shown to break down at next-to-next-to leading order [46]. Worse still, this ansatz gives the wrong sign for the ratio of the vector and the axial-vector dimension $d = 6$ condensates [47, 48], and it underestimates the $d = 6$ condensate in the vector channel by a factor 10 [47, 48].

This does not preclude attempts at determining the numerical values of vacuum condensates of higher dimensionality [17]. What would be lacking is their identification in terms of quark-gluon fields. Given the deterioration of uncertainties with increasing dimensionality, the determination of the dimension $d = 8$ condensate appears to be the borderline [17, 49, 50]. An example of the result for the dimension $d = 8$ chiral condensate $\langle \mathcal{O}_8 \rangle$ from [17] is shown below in Fig. 8.5.

References

1. M. A. Shifman, A. I. Vainshtein, V. I. Zakharov, Nucl. Phys. B **147**, 385 (1979). (ibid., B **147**, 448 (1979))
2. B. Guberina, R. Meckbach, R.D. Peccei, R. Rückl, Nucl. Phys. B **184**, 476 (1981)
3. J.S. Bell, R.A. Bertlmannn, Nucl. Phys. B **177**, 218 (1981)
4. J.S. Bell, R.A. Bertlmannn, Phys. Lett. B **137**, 107 (1984)
5. A. Bradley, C. S. Langensiepen, G. Shaw, Phys. Lett. B **102** 180, 359
6. D.J. Broadhurst, Phys. Lett. B **123**, 251 (1983)
7. J. Marrow, G. Shaw, Z. Phys, C-Particles and Fields **33**, 237 (1986)
8. J. Marrow, J. Parker, G. Shaw, Z. Phys. C-Particles and Fields **37**, 103 (1987)

9. S.I. Eidelman, L.M. Kurdadze, A.I. Vainshtein, Phys. Lett. B **82**, 278 (1979)
10. R.A. Bertlmann, C.A. Dominguez, M. Loewe, M. Perrotet, E. de Rafael, Z. Phys. C-Particles and Fields **39**, 231 (1988)
11. H. Albrecht et al., Z. Phys. C-Particles and Fields **33**, 7 (1986)
12. C.A. Dominguez, J. Solá, Z. Phys. C-Particles and Fields **40**, 63 (1988)
13. M. Beneke, M. Jamin, J. High Energy Phys. **01**, 125 (2013)
14. P. Minkowski, private communication
15. K.G. Patrignani et al., Particle Data Group. Chin. Phys. C **40**, 100001 (2016)
16. S. Bodenstein, C.A. Dominguez, S.I. Eidelman, H. Spiesberger, K. Schilcher, J. High Energy Phys. **01**, 039 (2012)
17. C.A. Dominguez, L.A. Hernandez, K. Schilcher, H. Spiesberger, J. High Energy Phys. **03**, 053 (2015)
18. C.A. Dominguez, L.A. Hernandez, K. Schilcher, J. High Energy Phys. **07**, 110 (2015)
19. S.G. Gorishnii, A.L. Kataev, S.A. Larin, Phys. Lett. B **259**, 144 (1991)
20. T. van Ritbergen, J.A.M. Vermaseren, S.A. Larin, Phys. Lett. B **400**, 379 (1997)
21. P.A. Baikov, K. Chetyrkin, J.H. Kühn, Phys. Rev. Lett. **101**, 012002 (2008)
22. K. Chetyrkin, B.A. Kniehl, M. Steinhauser, Phys. Rev. Lett. **79**, 2184 (1997)
23. M. Davier, A. Höcker, Z. Zhang, Rev. Mod. Phys. **78**, 1043 (2006)
24. A. Pich, Progr. Part. Nucl. Phys. **75**, 41 (2014)
25. D. Cronin-Hennessy et al., CLEO 2009. Phys. Rev. D **80**, 072001 (2009)
26. J.Z. Bai et al., BES 2006. Phys. Rev. Lett. **97**, 262001 (2006)
27. D.J. Broadhurst et al., Phys. Lett. B **329**, 103 (1994)
28. J.H. Kühn, M. Steinhauser, C. Sturm, Nucl. Phys. B **778**, 192 (2007)
29. R. Horsley et al., Phys. Rev. D **86**, 054502 (2012)
30. C. McNeile et al., PQCD Collaboration. Phys. Rev. D **82**, 034512 (2010)
31. S. Bodenstein, J. Bordes, C.A. Dominguez, J. Peñarrocha, K. Schilcher, Phys. Rev. D **83**, 074014 (2011)
32. R. Boughezal, M. Czakon, T. Schutzmeier, Phys. Rev. D **74**, 074006 (2006)
33. A. Maier, P. Maierhöfer, P. Marquard, Phys. Lett. B **669**, 88 (2008)
34. R. Boughezal, M. Czakon, T. Schutzmeier, Nucl. Phys. B (Proc. Suppl.) **160** (2006) 164
35. K.G. Chetyrkin, J.H. Kühn, C. Sturm, Eur. Phys. J. C **48**, 107 (2006)
36. A. Maier, P. Maierhöfer, P. Marquard, Nucl. Phys. B **797**, 218 (2008)
37. A. Maier, P. Maierhöfer, P. Marquard, A.V. Smirnov, Nucl. Phys. B **824**, 1 (2010)
38. K. Chetyrkin et al., Theor. Math. Phys. **170**, 217 (2012)
39. K.G. Chetyrkin, R. Harlander, J.H. Kühn, M. Steinhauser, Nucl. Phys. B **503**, 339 (1997)
40. P.A. Baikov, K. Chetyrkin, J.H. Kühn, Nucl. Phys. Proc. Suppl. B **135**, 243 (2004)
41. K.G. Chetyrkin et al., Phys. Rev. D **80**, 074010 (2009)
42. A. Hoang, M. Jamin, Phys. Lett. B **594**, 127 (2004)
43. J.Z. Bai et al., BES 2002. Phys. Rev. Lett. **88**, 101802 (2002)
44. R. Horsley et al., arXiv: 1205.1659
45. B.V. Geshkenbein, Phys. Rev. D **70**, 074027 (2004)
46. A. Gomez Nicola, J.R. Pelaez, J. Ruiz de Elvira, Phys. Rev. D **82**, 074012 (2010)
47. J. Bordes, C.A. Dominguez, J. Peñarrocha, K. Schilcher, J. High Energy Phys. **02**, 037 (2006)
48. C.A. Dominguez, K. Schilcher, J. High Energy Phys. **01**, 093 (2007)
49. M. Gonzalez-Alonso, A. Pich, J. Prades, Phys. Rev. D **81**, 074007 (2010)
50. M. Gonzalez-Alonso, A. Pich, J. Prades, Phys. Rev. D **82**, 014019 (2010)

Chapter 9
Quark Masses

Light quark masses (up-, down-, strange-) were first considered in the framework of chiral symmetries and named *current algebra quark masses* [1–5]. These quantities, being pre-QCD, lacked the current detailed understanding of e.g. quark-mass running and renormalization. The mechanism of global $SU(3) \times SU(3)$ chiral symmetry realized in the Nambu-Goldstone fashion, and its breaking down to $SU(2) \times SU(2)$, followed by a breaking down to $SU(2)$, and finally to $U(1)$ was first understood using the strong interaction Hamiltonian [2, 4, 5]

$$H(x) = H_0(x) + \epsilon_0\, u_0(x) + \epsilon_3\, u_3(x) + \epsilon_8\, u_8(x) \,. \tag{9.1}$$

The term $H_0(x)$ is $SU(3) \times SU(3)$ invariant, the $\epsilon_{0,3,8}$ are symmetry breaking parameters, and the scalar densities $u_{0,3,8}(x)$ transform according to the $3\bar{3} \oplus \bar{3}3$ representation of $SU(3) \times SU(3)$. Only ratios of these densities yield finite results. In QCD language, ϵ_8 is related to the strange quark mass m_s, and ϵ_3 to the quark mass difference $m_d - m_u$. The scalar densities are related to products of quark-anti-quark field operators. For instance, the ratio of $SU(3)$ to $SU(2)$ breaking is given by

$$R \equiv \frac{m_s - m_{ud}}{m_d - m_u} = \frac{\sqrt{3}}{2}\, \frac{\epsilon_8}{\epsilon_3} \,, \tag{9.2}$$

where $m_{ud} \equiv (m_u + m_d)/2$. Before QCD many relations for quark-mass ratios were obtained from hadron mass ratios, as well as from other hadronic information, e.g. $\eta \to 3\pi$, K_{l3} decay, etc. [5]. A pioneering determination of R, Eq. (9.2), from a solution to the $\eta \to 3\pi$ puzzle proposed in [6], was obtained in [7], $R^{-1} = 0.020 \pm 0.002$, in remarkable agreement with a later determination based on baryon mass splitting [8] $R^{-1} = 0.021 \pm 0.003$, and with the most recent value [9] $R^{-1} = 0.025 \pm 0.003$.

With the advent of CHPT [2, 5, 9, 10], certain quark mass ratios turned out to be renormalization scale independent to leading order, and could be expressed in terms

© The Author(s), under exclusive licence to Springer Nature Switzerland AG 2018
C. A. Dominguez, *Quantum Chromodynamics Sum Rules*,
SpringerBriefs in Physics, https://doi.org/10.1007/978-3-319-97722-5_9

of pseudoscalar meson mass ratios [5, 11, 12], e.g.

$$\frac{m_u}{m_d} = \frac{M_{K^+}^2 - M_{K^0}^2 + 2M_{\pi^0}^2 - M_{\pi^+}^2}{M_{K^0}^2 - M_{K^+}^2 + M_{\pi^+}^2} = 0.56 \,, \tag{9.3}$$

$$\frac{m_s}{m_d} = \frac{M_{K^+}^2 + M_{K^0}^2 - M_{\pi^+}^2}{M_{K^0}^2 - M_{K^+}^2 + M_{\pi^+}^2} = 20.2 \,, \tag{9.4}$$

where the numerical results follow after some subtle corrections due to electromagnetic self energies [9]. The current values of these ratios from the FLAG group are [9]

$$\frac{m_u}{m_d} = 0.46 \pm 0.03 \,, \tag{9.5}$$

$$\frac{m_s}{m_{ud}} = 27.43 \pm 0.31 \,. \tag{9.6}$$

Knowledge of these quark mass ratios is extremely important in the determination of individual values of light-quark masses from QCD sum rules. The reason being that the correlator of the axial-vector divergences, $\psi_5(q^2)$, Eq. (1.5), employed for this purpose involves as an overall multiplicative factor the terms $(m_u + m_d)$, or $(m_{ud} + m_s)$.

Beyond leading order in CHPT things become complicated. At next to leading order (NLO) the only parameter-free relation is

$$Q^2 \equiv \frac{m_s^2 - m_{ud}^2}{m_d^2 - m_u^2} = \frac{M_K^2 - M_\pi^2}{M_{K^0}^2 - M_{K^+}^2} \frac{M_K^2}{M_\pi^2} \,. \tag{9.7}$$

Other quark mass ratios at NLO and beyond depend on the renormalization scale, as well as on some CHPT low energy constants which need to be determined independently [9–11]. After taking into account electromagnetic self energies, Eq. (9.7) gives [11] $Q = 24.3$, a recent analysis of $\eta \to 3\pi$ [10, 11] gives $Q = 22.3 \pm 0.8$, and the current value from the FLAG Collaboration [9] is

$$Q = 22.6 \pm 0.7 \pm 0.6 \,. \tag{9.8}$$

The ratios R, Eq. (9.2), and Q, Eq. (9.7), together with the leading order ratios Eqs. (9.3)–(9.4), will prove useful for a comparisons with QCD sum rule results. An additional important quark mass ratio involving the ratios Eqs. (9.3)–(9.4) is

$$r_s \equiv \frac{m_s}{m_{ud}} = \frac{2\,m_s/m_d}{1 + m_u/m_d} = 28.1 \pm 1.3 \,, \tag{9.9}$$

where the numerical value follows from the NLO CHPT relation [11], to be compared with the LO result from Eqs. (9.3)–(9.4), $r_s = 25.9$, and a large N_c estimate [13]

$r_s = 26.6 \pm 1.6$. The most recent FLAG Collaboration result is [9]

$$r_s = 27.46 \pm 0.15 \pm 0.41 \, . \tag{9.10}$$

In order to go beyond light-quark mass ratios and determine analytically their individual values, including the charm and bottom quark masses, one needs suitable current correlators. In the light-quark sector (up-, down-, strange-quark) the appropriate correlator is that of the divergence of the axial-vector current $\psi_5(q^2)$, Eq. (1.5). This is quite convenient as its QCD expression involves the square of the quark masses as overall factor. For the heavy-quark masses the standard procedure uses the vector current correlator which involves these masses in both the high energy, as well as the low energy QCD expansions.

Starting with the light-quark masses, the case of the strange quark will be analysed in full detail following its latest determinations [14–16]. The two pieces of the FESR involving $\psi_5(q^2)$ are the QCD part

$$\delta_5^{QCD}(s_0) \equiv -\frac{1}{2\pi i} \oint_{C(|s_0|)} ds \, \psi_5^{QCD}(s) \, \Delta_5(s) \, , \tag{9.11}$$

and the hadronic part

$$\delta_5^{HAD}(s_0) \equiv 2 \, f_P^2 \, M_P^4 \, \Delta_5(M_P^2) + \int_{s_{th}}^{s_0} ds \, \frac{1}{\pi} \operatorname{Im} \psi_5(s)|_{RES} \, \Delta_5(s) \, , \tag{9.12}$$

where $\Delta_5(s)$ is an (analytic) integration kernel to be introduced shortly, the first term on the right hand side is the pseudoscalar meson pole contribution ($P = K$), s_{th} is the hadronic threshold, and $text Im \, \psi_5(s)|_{RES}$ is the hadronic resonance spectral function. The radius of integration s_0 is assumed to be large enough for QCD to be valid on the circle.

For later convenience the FESR, $\delta_5(s_0)|_{HAD} = \delta_5(s_0)|_{QCD}$ can be rewritten as

$$\delta_5(s_0)|^{QCD} = \delta_5|_{POLE} + \delta_5(s_0)|_{RES} \, , \tag{9.13}$$

where the meaning of each term is self evident. The full expression of ψ_5^{QCD} and its second derivative are given in Appendix **C**, while the contour integrals, $\delta_5^{QCD}(s_0)$, are given in Appendix **D**. The quark-mass squared $(\overline{m}_s + \overline{m}_q)^2$, with $\overline{m}_q = (\overline{m}_u + \overline{m}_d)/2$, enters in $\delta_5(s_0)^{QCD}$ as an overall factor. Hence, it is given by the ratio

$$(\overline{m}_s + \overline{m}_q)^2 = \frac{\delta_5(s_0)|_{HAD}}{\hat{\delta}_5(s_0)|_{QCD}} \tag{9.14}$$

where $\hat{\delta}_5(s_0)|_{QCD}$ obviously lacks the overall quark mass squared factor. The quark mass $\overline{m}_q$ stands for the running mass in the $\overline{MS}$ scheme. If one were to ignore the hadronic resonance contribution in Eq. (9.12) then the quark mass following from

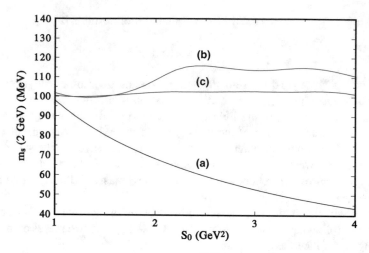

Fig. 9.1 The strange quark mass $\overline{m}_s$(2 GeV) in the $\overline{MS}$ scheme taking into account only the kaon pole with $\Delta_5(s) = 1$ (curve (**a**)), and the two Breit-Wigner resonance spectral function with a threshold constraint from CHPT, Eq. (6.5), with $\Delta_5(s) = 1$ (curve (**b**)), and $\Delta_5(s)$ as in Eq. (9.15) (curve (**c**)). A systematic uncertainty of some 20 % due to the resonance sector is dramatically unveiled (difference between curves (**b**) and (**c**))

Eq. (9.14) would be a monotonically decreasing function of s_0, thus providing no information. This is illustrated in Fig. (9.1) for $\Delta_5(s) = 1$ (curve(a)).

Turning to the integration kernel, $\Delta_5(s)$, entering the FESR its purpose is to quench the contribution of the hadronic resonances beyond the pole. Historically, these resonances were predicted long ago [17, 18] as radial excitations of the pseudoscalar ground states, i.e. the pion and the kaon. The motivation in [17] was to account for chiral symmetry breaking corrections in $SU(2) \times SU(2)$ (pion case) and $SU(3) \times SU(3)$ (kaon case). The predicted masses of these resonances from the dual-resonance model agreed well with the experimental values [19], once these states were observed years later in hadronic interactions. However, information on the existence and on the mass and width of these resonances is hardly enough to reconstruct the hadronic spectral function. For instance, potential non-resonant background, inelasticity and resonance interference are impossible to predict. Hence, while their presence is required, the lack of relevant information introduces a systematic uncertainty in the results. In order to try and quench this uncertainty the integration kernel $\Delta_5(s)$ is introduced such that e.g. it vanishes at the peak of whatever functional form is chosen for the resonant states, e.g. a Breit-Wigner form. Assuming two radial excitations the kernel becomes a quadratic function of the squared energy,s, i.e.

$$\Delta_5(s) = 1 - a_0\, s - a_1\, s^2 \,, \tag{9.15}$$

where the constant coefficients a_0 and a_1 are chosen such that

$$\Delta_5(M_1^2) = \Delta_5(M_2^2) = 0 \tag{9.16}$$

where $M_{1,2}$ are the masses of the resonances. An equivalent expression is given by [16]

$$\Delta_5(s) = (s - M_1^2)(s - M_2^2), \tag{9.17}$$

and two other kernels were considered in [16]

$$\Delta_5(s) = (s - M_1^2)(s - M_2^2)(s_0 - s), \tag{9.18}$$

and

$$\Delta_5(s) = (s - a)(s - s_0), \tag{9.19}$$

where the free parameter a was determined by demanding maximal reduction of the uncertainty in the quark mass.

Regarding the resonance spectral functions, historically they were initially parametrized by Breit-Wigner forms, albeit without any specific threshold normalization. A procedure to normalize at threshold according to chiral perturbation theory was eventually introduced in [20], thus reducing this systematic uncertainty. In the case of the kaon channel the threshold behaviour is described in detail in Chap. 6, with results given in Eqs. (6.11)–(6.13). It is important to notice the importance of the sub-channel $K^*(892) - \pi$. Due to the narrow width of the $K^*(892)$ this contribution is not negligible. The two resonances entering the integration kernel $\Delta_5(s)$ are the K(1460) and K(1830), both with widths of 250 MeV.

The crucial importance of the kernel, Eq. (9.15), is highlighted in Fig. 9.1. Curve (b) is the result for the strange-quark mass with no integration kernel, i.e. with $\Delta_5(s) = 1$, in which case both resonances contribute 100% to the mass. Given the stability of the result one would have determined $\overline{m}_s(2\,\text{GeV}) \simeq 100 - 120\,\text{MeV}$, with an unknown systematic uncertainty from the hadronic sector. The presence of the integration kernel leads to curve (c), not only a more stable prediction, but one unveiling the systematic uncertainty, and a result some 20% lower.

Turning to Eq. (9.14), in order to determine $\overline{m}_s$ one needs information on the quark mass ratio $\overline{m}_q/\overline{m}_s$, which is provided by e.g. chiral perturbation theory. All the ingredients up to this point were used in [15] to find

$$\overline{m}_s(2\,\text{GeV}) = \begin{cases} 95 \ \pm 5 \ \text{MeV} \\ 111 \ \pm 6 \ \text{MeV}, \end{cases} \tag{9.20}$$

corresponding to $\Lambda_{QCD} = 420\ (330)$ MeV, respectively. While several sources of uncertainty were considered in [15] to arrive at this result, including a conservative guess of the unknown six-loop perturbative QCD contribution, the convergence of the perturbative expansion was not explicitly taken into account. Furthermore, the use of the strong running coupling α_s involving the scale Λ_{QCD}, Eq. (3.10), is responsible for a rather large uncertainty in the quark mass. Theses issues were addressed in

[14, 16] as follows. The strong running coupling, α_s, was determined using Eq. (3.9) together with its precise value at the Z-boson scale, scaled down to $\alpha_s(s_0)$. The various integration kernels, Eqs. (9.15)–(9.19) were used with the results contributing to the final uncertainty. Of essential importance, an analysis of the convergence of the perturbative QCD (PQCD) expansion was performed as follows. Numerically, the result for δ_5^{PQCD} using the integration kernel Eq. (9.17) with $s_0 = 4.2\,\mathrm{GeV}^2$, and the renormalization scale $\mu = \sqrt{s_0}$, is given by the expansion

$$\delta_5^{PQCD} = 0.23\,\mathrm{GeV}^8\left[1 + 2.2\,\alpha_s + 6.7\,\alpha_s^2 + 19.5\,\alpha_s^3 + 56.5\,\alpha_s^4\right], \qquad (9.21)$$

which after replacing a typical value of α_s leads to all terms beyond the leading order to be roughly the same, e.g. for $\alpha_s = 0.3$ the result is

$$\delta_5^{PQCD} = 0.23\,\mathrm{GeV}^8\left[1 + 0.65 + 0.60 + 0.53 + 0.46\right], \qquad (9.22)$$

which does not look at all convergent. To judge from the first five terms, this expansion is worse behaved than the non-convergent harmonic series. Since the quark mass is actually proportional to the inverse square-root of δ_5^{QCD}, after expanding the latter gives instead

$$(\delta_5^{PQCD})^{-1/2} = 2.08\,\mathrm{GeV}^{-4}\left[1 - 1.10\,\alpha_s - 1.52\,\alpha_s^2 - 2.08\,\alpha_s^3 - 3.21\,\alpha_s^4\right], \qquad (9.23)$$

which exhibits a considerably improved convergence, e.g. for $\alpha_s = 0.3$ this expansion becomes

$$(\delta_5^{PQCD})^{-1/2} = 2.08\,\mathrm{GeV}^{-5}\left[1 - 0.33 - 0.14 - 0.06 - 0.03\right]. \qquad (9.24)$$

It is interesting to notice that this expansion of the inverse square-root is equivalent to the lowest Padè approximant.

A thorough analysis of the various sources of uncertainty can be found in the original reference [16]. The final result for the strange-quark mass at a scale $\mu = 2\,\mathrm{GeV}$ is

$$\overline{m}_s(2\,\mathrm{GeV}) = 94 \pm 8\,\mathrm{MeV}. \qquad (9.25)$$

Turning to the up- and down-quark masses were determined in [21] to five-loop order in PQCD, and using a quadratic integration kernel of the form Eq. (9.15). However, the convergence of the PQCD expansion was not analysed as above for the strange quark mass. This is currently an ongoing project being finalized [22].

Turning to the heavy-quark sector, the charm-quark mass determination [23] is discussed next (the bottom-quark case is treated similarly in [24]). The starting point is the choice of the heavy-quark vector current correlator

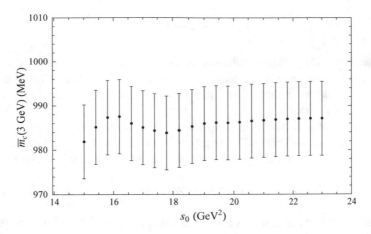

Fig. 9.2 The charm-quark mass at a scale $\mu = 3\,\mathrm{GeV}$, as a function of s_0, for the kernel Eq. (9.30)

$$\Pi_{\mu\nu}(q^2) = i \int d^4x \; e^{iqx} \langle 0|T(V_\mu(x) \; V_\nu(0))|0\rangle$$
$$= (q_\mu \, q_\nu - q^2 g_{\mu\nu}) \, \Pi_V(q^2) \,, \qquad (9.26)$$

where $V_\mu(x) = \bar{c}(x)\gamma_\mu c(x)$.

For reasons to become clear later the chosen FESR involves an meromorphic integration kernel, $p(s)$, so that the residue theorem in the complex s-plane is written as

$$\int_0^{s_0} p(s)\, \frac{1}{\pi} Im\, \Pi_V(s)\, ds = -\frac{1}{2\pi i} \oint_{C(|s_0|)} p(s)\, \Pi_V(s)\, ds$$
$$+ \mathrm{Res}[\Pi_V(s)\, p(s), s = 0] \,, \qquad (9.27)$$

where

$$Im\, \Pi_V(s) = \frac{1}{12\pi}\, R_c(s) \,, \qquad (9.28)$$

with $R_c(s)$ the standard R-ratio for charm production. The low energy expansion of $\Pi_V(s)$ around $s = 0$ is given in Eq. (8.37), and the coefficients $\overline{C}_n$ can be expanded in powers of $\alpha_s(\mu)$ as in Eq. (8.38). At short distances the perturbative expansion of $\Pi_V(s)$ is given in Eqs. (8.40)–(8.41). The experimental data for $R_c(s)$, used in the determination of the gluon condensate had been used in this application. Finally, two integration kernels were chosen in [23]

$$p_1(s) = \frac{1}{s^2} \,, \qquad (9.29)$$

and

$$p_2(s) = 1 - \left(\frac{s_0}{s}\right)^2,$$ (9.30)

Given that this correlation function was used in the determination of the gluon condensate in Chap. 8, it is important to mention that the contribution of this condensate in the determination of the charm-quark mass is negligible. In other words there is no *bootstrap*.

It turns out that kernel $p_2(s)$, Eq. (9.30) leads to a far more precise value of the charm-quark mass (Fig. 9.2). Using this kernel, and after taking into account all possible sources of uncertainty the quark mass at a scale of 3 GeV becomes

$$\overline{m}_c(3\,\text{GeV}) = 987 \pm 9\,\text{MeV},$$ (9.31)

where this scale is chosen to allow for a direct comparison with LQCD results [9], with which there is good agreement. This result agrees within errors with a determination using a different integration kernel [25]. For other determinations see [26–29].

The determination of the bottom quark mass [24] follows closely that of the charm quark so it will not be discussed here. The result is

$$\overline{m}_b(10\,\text{GeV}) = 3623 \pm 9\,\text{MeV},$$ (9.32)

and

$$\overline{m}_b(\overline{m}_b) = 4171 \pm 9\,\text{MeV},$$ (9.33)

in good agreement with LQCD [30].

References

1. S. B. Treiman, R. Jackiw, B. Zumino, E. Witten, *Current Algebra and Anomalies* (World Scientific, Singapore, 1985). See also: J. Bernstein, *Elementary Particles and their Currents* (W. H. Freeman & Co., San Francisco, 1968)
2. H. Pagels, Phys. Rep. **16**, 219 (1975)
3. M. Gell-Mann, R.J. Oakes, B. Renner, Phys. Rev. **175**, 2195 (1968)
4. S. Glashow, S. Weinberg, Phys. Rev. Lett. **20**, 224 (1968)
5. J. Gasser, H. Leutwyler, Phys. Rep. C **87**, 77 (1982)
6. C.A. Dominguez, A. Zepeda, Phys. Rev. D **18**, 884 (1978)
7. C.A. Dominguez, Phys. Lett. B **86**, 171 (1979)
8. P. Minkowski, A. Zepeda, Nucl. Phys. B **164**, 25 (1980)
9. S. Aoki et al., Eur. Phys. J. C **77**, 112 (2017)
10. G. Colangelo, S. Lanz, E. Passemar, PoS CD **09**, 047 (2009)
11. H. Leutwyler, PoS CD **09**, 005 (2009)
12. S. Weinberg, Trans. New York Acad. Sci. **38**, 185 (1977)
13. H. Leutwyler, Nucl. Phys. B Proc. Suppl. **64**, 223 (1998)
14. C. A. Dominguez, Analytical determination of QCD quark masses, in *Fifty Years of Quarks*, ed. by H. Fritzsch, M. Gell-Mann (World Scientific Publishing Co., Singapore, 2015), pp. 287–313
15. C.A. Dominguez, N.F. Nasrallah, R. Röntsch, K. Schilcher, J. High Energy Phys. **05**, 020 (2008)

16. S. Bodenstein, C.A. Dominguez, K. Schilcher, J. High Energy Phys. **07**, 138 (2013)
17. C. A. Dominguez, Phys. Rev. D **7**, 1252 (1973); *ibid.* D **16**, 2320 (1977); Riv. Nuovo Cim. **8N6**, 1 (1985)
18. P. Frampton, Phys. Rev. D **1**, 3141 (1970)
19. K.G. Patrignani et al., Particle data group. Chin. Phys. C **40**, 100001 (2016)
20. C. A. Dominguez, Z. Phys. C **26**, 269 (1984)
21. C.A. Dominguez, N.F. Nasrallah, R.H. Rontsch, K. Schilcher, Phys. Rev. D **79**, 014009 (2009)
22. C. A. Dominguez, A. Mes, K. Schilcher, to be published
23. S. Bodenstein, J. Bordes, C.A. Dominguez, J. Peñarrocha, K. Schilcher, Phys. Rev. D **83**, 074014 (2011)
24. S. Bodenstein, J. Bordes, C.A. Dominguez, J. Peñarrocha, K. Schilcher, Phys. Rev. D **85**, 034003 (2012)
25. S. Bodenstein, J. Bordes, C.A. Dominguez, J. Peñarrocha, K. Schilcher, Phys. Rev. D **82**, 114013 (2010)
26. J.H. Kühn, M. Steinhauser, C. Sturm, Nucl. Phys. B **778**, 192 (2007)
27. K.G. Chetyrkin et al., Phys. Rev. D **80**, 074010 (2009)
28. B. Dehnadi, A.H. Hoang, V. Mateu, S.M. Zerbarjard, *American Institute of Physics Conference Proceedings*, vol. 1441 (2012), p. 628
29. A. Signer, Phys. Lett. B **672**, 333 (2009)
30. C. McNeile et al., PQCD collaboration. Phys. Rev. D **82**, 034512 (2010)

Chapter 10
Corrections to the GMOR Relation

The Gell-Mann-Oakes-Renner (GMOR) relation was introduced in the Introduction, Chap. 1, Eq. (1.14). In this section it will be discussed in more detail. While originally obtained in the framework of current algebra and chiral symmetry, it is a low energy theorem of QCD. It involves the correlator of the divergence of the QCD axial-vector current, Eq. (1.7), leading to the low energy theorems

$$\psi_5(0)|_u^d = -(m_u + m_d)\langle 0|\bar{u}u + \bar{d}d|0\rangle + \mathcal{O}(m_{u,d}^2), \qquad (10.1)$$

in $SU(2) \times SU(2)$, and

$$\psi_5(0)|_s^q = -(m_s + m_q)\langle 0|\bar{s}s + \bar{q}q|0\rangle + \mathcal{O}(m_s^2), \qquad (10.2)$$

in $SU(3) \times SU(3)$, where $m_q \equiv (m_u + m_d)/2$. Considering the correlator $\psi_5(s)$ in the complex squared energy s-plane, Fig. 10.1, from Cauchy's theorem one has

$$\psi_5(0) = \frac{1}{2\pi i} \oint \frac{\psi_5(s)}{s}\, ds. \qquad (10.3)$$

In the hadronic sector there is a pole on the real axis corresponding to the pion or the kaon, followed by higher resonances

$$\psi_5(s)|_{HAD} = \frac{2 f_P^2 M_P^4}{M_P - s}, + \psi_5(s)|_{Res}, \qquad (10.4)$$

where f_P and M_P are respectively the decay constant and the mass of the pion or kaon, with $f_K/f_\pi = 1.197 \pm 0.006$, $f_\pi = 92.21 \pm 0.14$ MeV [1]. Equation (10.3) then becomes

$$\psi_5(0)|_i^j \equiv -(m_i + m_j)\,\langle 0|\,\bar{q}_i q_i + \bar{q}_j q_j\,|0\rangle = 2 f_P^2 M_P^2 (1 - \delta_P), \qquad (10.5)$$

where i, j stand for the corresponding quark flavours, and δ_P is the (hadronic) correction from the resonance contribution. This correction is rather important as it is

© The Author(s), under exclusive licence to Springer Nature Switzerland AG 2018
C. A. Dominguez, *Quantum Chromodynamics Sum Rules*,
SpringerBriefs in Physics, https://doi.org/10.1007/978-3-319-97722-5_10

related to two of chiral perturbation theory (CHPT) low energy constants, L_8^r and H_2^r, i.e.

$$\delta_\pi = 4\frac{M_\pi^2}{f_\pi^2}(2L_8^r - H_2^r) \quad \text{and} \quad \delta_K = \frac{M_K^2}{M_\pi^2}\delta_\pi. \qquad (10.6)$$

The low energy constants L_8^r and H_2^r also enter in the ratio $\langle \bar{s} s\rangle/\langle \bar{q} q\rangle$ through

$$R_{sq} \equiv \frac{<\bar{s}s>}{<\bar{q}q>} = 1 + 3\mu_\pi - 2\mu_K - \mu_\eta + \frac{8}{f_\pi^2}(M_K^2 - M_\pi^2)(2L_8^r + H_2^r), \qquad (10.7)$$

where

$$\mu_P = \frac{M_P^2}{32\pi^2 f_\pi^2} \ln \frac{M_P^2}{v_\chi^2}, \qquad (10.8)$$

with v_χ the chiral renormalization scale.

In the past there was some debate on the feasibility of determining the hadronic corrections δ_π and δ_K using QCDSR. This issue was fully resolved in the affirmative in [2–4].

The determination of the hadronic corrections δ_π and δ_K involves Eq. (10.3). Introducing an analytic kernel $\Delta_5(s)$, and splitting the integration into a line integral along the positive real semi-axis, and a contour integral around a circle of radius $|s_0|$, gives

$$\psi_5(0)\Delta_5(0) = \frac{1}{\pi}\int_{S_{th}}^{s_0}\frac{\Delta_5(s)}{s}\,\text{Im}\,\psi_5(s)\,ds + \frac{1}{2\pi i}\oint_{C(|s_0|)}\frac{\Delta_5(s)}{s}\psi_5(s)\,ds, \qquad (10.9)$$

The integration kernel in the pionic channel was introduced in Eqs. (9.15)–(9.16), and the threshold behaviour in the pionic channel is given in Eq. (6.2), while for the kaon channel it is given in Eqs. (6.11)–(6.13) (Fig. 10.2).

A detailed analysis of the determination of δ_π is given in [5], and the case of δ_K is discussed in [6]. The results for $\psi_5(0)$ in both channels are very stable in the wide range $s_0 \simeq 3.0 - 5.0\,\text{GeV}^2$ leading to

$$\delta_\pi = (6.2 \pm 1.6)\%, \qquad (10.10)$$

$$\delta_K = (55 \pm 5)\%, \qquad (10.11)$$

corresponding to

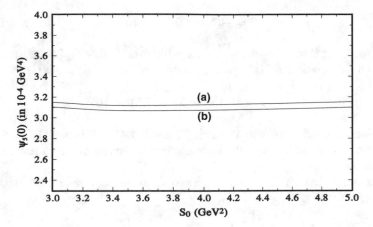

Fig. 10.1 Results for $\psi_5(0)$ in $SU(2) \times SU(2)$ in units of 10^{-4} GeV^4 as a function of s_0 and using a two-resonance parametrization. Curve (**a**) corresponds to $\alpha_s(M_\tau^2) = 0.335$ ($\Lambda = 365$ MeV), and curve (**b**) to $\alpha_s(M_\tau^2) = 0.353$ ($\Lambda = 397$ MeV)

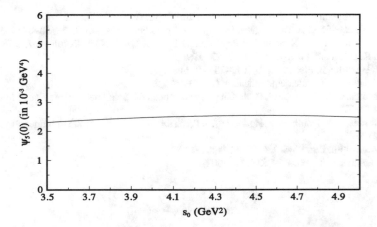

Fig. 10.2 Results for $\psi_5(0)$ in $SU(3) \times SU(3)$ in units of 10^{-3} GeV^4 as a function of s_0, and using a two-resonance parametrization. Curve (**a**) corresponds to $\alpha_s(M_\tau^2) = 0.335$ ($\Lambda = 365$ MeV), and curve (**b**) to $\alpha_s(M_\tau^2) = 0.353$ ($\Lambda = 397$ MeV)

$$\psi_5(0)|_u^d = (3.11 \pm 0.04) \times 10^{-4} \, \text{GeV}^4, \tag{10.12}$$

$$\psi_5(0)|_s^{ud} = (2.8 \pm 0.3) \times 10^{-3} \, \text{GeV}^4. \tag{10.13}$$

These results together with Eqs. (10.6)–(10.7) determine the low energy constants L_8^r and H_2^r

$$L_8^r = (1.0 \pm 0.3) \times 10^{-3}, \tag{10.14}$$

$$H_2^r = -(4.7 \pm 0.6) \times 10^{-3}. \tag{10.15}$$

The result for L_8^r is in very good agreement with several Lattice QCD determinations [7], and the unphysical H_2^r is in good agreement with the estimate [8]

$$H_2^r(\nu_\chi = M_\rho) = -(3.4 \pm 1.5) \times 10^{-3}. \tag{10.16}$$

Regarding the quark-mass corrections indicated in Eqs. (10.1) and (10.2), they are negligible in comparison with the hadronic corrections so that they can be safely ignored.

For a more thorough discussion of this subject the reader is referred to the original references [5, 6].

References

1. K.G. Patrignani et al., Particle data group. Chin. Phys. C **40**, 100001 (2016)
2. K.G. Chetyrkin, C.A. Dominguez, D. Pirjol, K. Schilcher, Phys. Rev. D **51**, 5090 (1995); K.G. Chetyrkin, D. Pirjol, K. Schilcher, Phys. Lett. B **404**, 337 (1997). C.A. Dominguez, L. Pirovano, and K. Schilcher. Phys. Lett. B **425**, 193 (1998)
3. D.J. Broadhurst, Phys. Lett. B **101**, 423 (1981)
4. H. Leutwyler, Private communication (2013)
5. J. Bordes, C.A. Dominguez, P. Moodley, J. Penarrocha, K. Schilcher, J. High Ener. Phys. **1005**, 064 (2010)
6. J. Bordes, C.A. Dominguez, P. Moodley, J. Penarrocha, K. Schilcher, J. High Ener. Phys. **1210**, 102 (2012)
7. S. Aoki et al., Eur. Phys. J. C **77**, 112 (2017)
8. M. Jamin, Phys. Lett. B bf **538**, 71 (2002)

Chapter 11
Anomalous Magnetic Moment of the Muon

The anomalous magnetic moment of the muon, due to quantum corrections, is related to the g-factor as

$$a_\mu \equiv \frac{g-2}{2} \neq 0. \tag{11.1}$$

The theoretical prediction of the muon anomaly, a_μ, currently disagrees with experiment at the level of some 3σ. The largest uncertainty in the Standard Model prediction comes from the leading order hadronic contribution, $a_\mu^{\mathrm{HAD,LO}}$ [1], to be discussed in this chapter. The traditional procedure to determine this quantity involves the $e^+e^- \rightarrow hadrons$ data, leading to [2]

$$a_\mu^{\mathrm{HAD}} = (693.1 \pm 3.4) \times 10^{-10}. \tag{11.2}$$

After adding the rest of the contributions to the anomaly, leading to the Standard Model (SM) result, a_μ^{SM}, the difference with the direct measurement, a_μ^{EXP}, is

$$a_\mu^{\mathrm{EXP}} - a_\mu^{\mathrm{SM}} = (26.8 \pm 7.6) \times 10^{-10}, \tag{11.3}$$

a 3.5σ discrepancy. This result has prompted a plethora of attempts over the years to explain this discrepancy beyond the SM. Before entertaining such an extreme possibility one should attempt to predict the anomaly entirely from theory, i.e. QCD. This proposition is also supported by the fact that the e^+e^- data base leading to Eq. (11.3) comprises many different experiments, at various facilities, performed over the years, and involving a large number of different final states. For a detailed and critical analysis of this data base see [3].

An attempt to account for the hadronic muon anomaly entirely within QCD was made in [4, 5] and discussed in the sequel.

The standard expression for the lowest order muon anomaly is given by

© The Author(s), under exclusive licence to Springer Nature Switzerland AG 2018
C. A. Dominguez, *Quantum Chromodynamics Sum Rules*,
SpringerBriefs in Physics, https://doi.org/10.1007/978-3-319-97722-5_11

$$a_\mu^{\text{HAD}} = \frac{\alpha_{EM}^2}{3\pi^2} \int_{s_{th}}^{\infty} \frac{ds}{s} K(s) R(s),$$ (11.4)

where α_{EM} is the electromagnetic coupling, and the standard R-ratio is

$$R(s) = 3 \sum_f Q_f^2 \left[8\pi \, \text{Im} \, \Pi(s) \right],$$ (11.5)

where Q_f are the quark charges and $\Pi(s)$ is the vector current correlator, Eq. (8.6), normalized to $8\pi \, \text{Im} \, \Pi(s) = [1 + \alpha_s/\pi + \cdots]$ (notice dropped upper label EM). The integration kernel $K(s)$ in Eq. (11.4) is given by [6]

$$K(s) = \int_0^1 dx \, \frac{x^2(1-x)}{x^2 + \frac{s}{m_\mu^2}(1-x)},$$ (11.6)

where m_μ is the muon mass. The traditional approach to determine a_μ^{HAD} is to split the integration region in Eq. (11.6) into a low-energy piece from threshold up to $s = s_0 \simeq (1.8\,\text{GeV})^2$, and a a high energy contribution from $s = s_0$ to infinity. The low energy integral involves the data (e^+e^- annihilation or tau-lepton decay into hadrons), while the high energy integral is computed in perturbative QCD (PQCD). For convenience it is useful to split the contributions to the integral in Eq. (11.4) into the three main regions, dominated by the light-, charm- and bottom-quark

$$a_\mu^{HAD} = a_\mu^{HAD}|_{uds} + a_\mu^{HAD}|_c + a_\mu^{HAD}|_b.$$ (11.7)

While the integration kernel $K(s)$ is known, and given in Eq. (11.6), a fit to it in terms of a meromorphic function allows using the power of the complex squared energy plane to determine the anomaly entirely from QCD. For this to rival the data driven determinations, the fit function must be extremely accurate. This is indeed possible thanks to the shape of $K(s)$, to wit.

Starting with the light-quark sector, the optimal fit function $K_1(s)$ is given by

$$K_1(s) = 2.257 \times 10^{-5}s + 3.482 \times 10^{-3}s^{-1}$$
$$- 1.467 \times 10^{-4}s^{-2} + 4.722 \times 10^{-6}s^{-3},$$ (11.8)

where s is expressed in GeV^2, and the numerical coefficients have the appropriate units to render $K_1(s)$ dimensionless. The functions $K(s)$ and $K_1(s)$ are shown in Fig. 11.1. The difference between the two functions is less than 0.08 % in the whole s-range!.

Next, in the charm-quark region $s_1 \simeq M_{J/\psi}^2 \leq s \leq s_2 \simeq (5.0\,\text{GeV})^2$, the optimal fit function $K_2(s)$ is

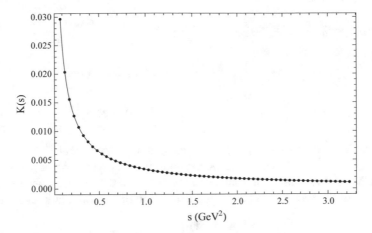

Fig. 11.1 The exact kernel $K(s)$, Eq. (11.6) (solid line) together with the fit in the light-quark region $K_1(s)$, Eq. (11.8), (solid circles). Relative difference between the two is less than 0.08%

$$K_2(s) = \frac{a_1}{s} + \frac{a_2}{s^2}, \tag{11.9}$$

where $a_1 = 0.003712$ GeV^2 and $a_2 = -0.0005122$ GeV^4, providing an excellent fit, with $K_2(s)$ differing from the exact kernel $K(s)$ by less than 0.02% in the whole range. For the bottom-quark region the fit function is given by

$$K_3(s) = 0.003719 \, \text{GeV}^2 \, s^{-1} - 0.0007637 \, \text{GeV}^4 \, s^{-2}, \tag{11.10}$$

differing from the exact kernel, $K(s)$, by less than 0.0005 % in the range $M_\Upsilon^2 \le s \le (12 \, \text{GeV})^2$.

The QCD sum rule in the light-quark region is given by

$$\int_{S_{th}}^{s_0} \frac{ds}{s} K_1(s) \ \frac{1}{\pi} \, \text{Im} \, \Pi_{uds}(s) = \text{Res} \left[\Pi_{uds}(s) \frac{K_1(s)}{s} \right]_{s=0}$$
$$- \frac{1}{2\pi i} \oint_{|s|=s_0} \frac{ds}{s} K_1(s) \, \Pi_{uds}(s), \tag{11.11}$$

where the contour integral around the circle of radius $s_0 \simeq (1.8 \, \text{GeV})^2$, is computed using PQCD. This is known up to five-loop level, Eq. (8.14). The contour integration can be performed using fixed order perturbation theory (FOPT) or, alternatively, contour improved perturbation theory (CIPT). In the present case the difference between the two integration methods turns out to be negligible, as discussed later. The residues are given in terms of derivatives of the correlator at zero momentum, which in principle can be determined in LQCD [5]. Hence Eq. (11.4) in this sector becomes

$$a_\mu^{HAD}|_{uds} = 8\alpha_{EM}^2 \sum_{i=u,d,s} Q_i^2 \left\{ \mathrm{Res} \left[\Pi_{uds}(s) \frac{K_1(s)}{s} \right]_{s=0} \right.$$

$$- \frac{1}{2\pi i} \oint_{|s|=s_0} \frac{ds}{s} K_1(s) \, \Pi_{uds}(s)|_{\mathrm{PQCD}}$$

$$\left. + \int_{s_0}^\infty \frac{ds}{s} K(s) \frac{1}{\pi} \mathrm{Im} \, \Pi_{uds}(s)|_{\mathrm{PQCD}} \right\}, \qquad (11.12)$$

where the last integral above involves the exact integration kernel $K(s)$. Analogous expressions follow in the charm-quark and bottom-quark sector. An important difference is that in these cases the residues can be obtained directly from the low energy expansion of the vector correlator [4]. The result in the charm-quark region is

$$\mathrm{Res} \left[\Pi_c(s)|_{PQCD} \frac{K_2(s)}{s} \right]_{s=0} = 76.1(5) \times 10^{-7}, \qquad (11.13)$$

where the error is due to the uncertainty in α_s and to the truncation of PQCD. In the bottom-quark region the residue is

$$\mathrm{Res} \left[\Pi_b(s)|_{PQCD} \frac{K_3(s)}{s} \right]_{s=0} = 6.3 \times 10^{-7}, \qquad (11.14)$$

where the error is negligible.

Turning to the integrals in Eq. (11.12) the contour integrals in fixed order perturbation theory and in the three regions are

$$\frac{1}{2\pi i} \oint_{|s|=s_0} \frac{ds}{s} K_n(s) \, \Pi_q(s)|_{\mathrm{PQCD}} = \begin{cases} 135.3(6) \times 10^{-7} \\ 20.3(1) \times 10^{-7} \\ 3.6(2) \times 10^{-7}, \end{cases} \qquad (11.15)$$

for $n = 1, 2, 3$ and $q = uds, c, b$, respectively. The result from integrating in contour improved perturbation theory differs by 0.2%. A comparison of these results with those from using e^+e^- annihilation data from the BES Collaboration [7] shows excellent agreement.

The line integrals in Eq. (11.12) in the three regions are

$$\int_{s_j}^\infty \frac{ds}{s} K(s) \frac{1}{\pi} \mathrm{Im} \, \Pi_q(s)|_{PQCD} = \begin{cases} 151.8(1) \times 10^{-7} \\ 20.0(4) \times 10^{-7} \\ 3.4(2) \times 10^{-7} \end{cases} \qquad (11.16)$$

with $j = 0, 2, 4$ corresponding to the regions $q = uds, c, b$, respectively. Substituting the numerical values of these two integrals into Eq. (11.12), and into its corresponding expressions for the charm- and bottom-quark regions, the leading order hadronic anomaly becomes

$$a_\mu^{HAD} = \frac{16}{3}\alpha_{EM}^2 \text{Res}\left[\Pi_{uds}(s)\frac{K_1(s)}{s}\right]_{s=0} + 19.4(2) \times 10^{-10}, \qquad (11.17)$$

where the Cauchy residue in the light-quark sector is discussed below. The individual contributions in the charm- and bottom-quark regions are fully determined after using Eqs. (11.13)–(11.14) with the results

$$a_\mu^{HAD}|_c = 14.4(1) \times 10^{-10}, \qquad (11.18)$$

$$a_\mu^{HAD}|_b = 0.29(1) \times 10^{-10}. \qquad (11.19)$$

A crucial vindication of this approach was provided later by LQCD determinations of these heavy-quark contributions [8, 9]

$$a_\mu^{HAD}|_c = 14.42(39) \times 10^{-10}, \qquad (11.20)$$

$$a_\mu^{HAD}|_b = 0.271(37) \times 10^{-10}, \qquad (11.21)$$

in excellent agreement with Eqs. (11.18)–(11.19).

Turning to the Cauchy residue in Eq. (11.12), given the expression of $K_1(s)$, Eq. (11.8), it can be written as

$$\text{Res}\left[\Pi_{uds}(s)\frac{K_1(s)}{s}\right]_{s=0} = \lim_{s\to 0}\sum_{n=1}^{3}\frac{a_n}{n!}\frac{d^n}{ds^n}\Pi_{uds}(s), \qquad (11.22)$$

where the a_n are the coefficients of the s^{-1}, s^{-2} and s^{-3} terms in Eq. (11.8), respectively. It is important to notice that the term proportional to s^{-1} in $K_1(s)$, Eq. (11.8), is positive, while that proportional to s^{-2} is negative. Hence, there is a partial cancellation among the first two terms in the residue, Eq. (11.22). The first two derivatives can be calculated in Lattice QCD from the slope of the vector current correlator. Results from the first determination are [10]

$$\frac{d}{dq^2}\Pi(s)_{uds}|_{s=0} = 0.07190 \pm 0.0025 \text{ GeV}^{-2}, \qquad (11.23)$$

$$\frac{d^2}{(dq^2)^2}\Pi(s)_{uds}|_{s=0} = 0.136 \pm 0.0009 \text{ GeV}^{-4} \qquad (11.24)$$

where these values correspond to the definition and normalization of the correlator as in Eq. (11.5), which differs from that in [10] by a factor 3/4. After substituting these results into Eq. (11.22) the residue becomes

$$\text{Res}\left[\Pi_{uds}(s)\frac{K_1(s)}{s}\right]_{s=0} = (0.240 \pm 0.009) \times 10^{-3} \tag{11.25}$$

so that the total value of the anomaly is

$$a_\mu^{HAD} = (701 \pm 26) \times 10^{-10}, \tag{11.26}$$

a result in need of considerable improvement, which should eventually be achieved in the future. In the meantime the residue can be obtained from the leading order hadronic saturation of the vector correlator, plus the Vector Meson Dominance (VMD) approximation to the pion form factor, as follows

$$\Pi_{\mu\nu}(q^2) = i \int d^4x e^{iqx} \int \frac{d^3p}{2\,p_0\,(2\pi)^3}\Big[\langle 0|V_\mu(x)|\pi\pi\rangle\langle|\pi\pi|V_\nu^\dagger(0)|0\rangle$$

$$+ \langle 0|V_\mu(x)|\rho\rangle\langle|\rho|V_\nu^\dagger(0)|0\rangle \cdots\Big], \tag{11.27}$$

The neglected terms correspond to multi-pion states. The first term above is loop-suppressed with respect to the second term, i.e. the single ρ-meson Born term. The matrix element in the second term above defines the ρ-photon coupling constant f_ρ according to

$$\langle 0|V_\mu(0)|\rho(p,s)\rangle = \frac{M_\rho^2}{f_\rho}\epsilon_\mu \tag{11.28}$$

where ϵ_μ is the ρ-meson polarization vector which satisfies the completeness relation

$$\sum_s \epsilon_\mu(p,s)\,\epsilon_\nu(p,s) = -g_{\mu\nu} + \frac{p_\mu\,p_\nu}{M_\rho^2} \equiv \Delta_{\mu\nu}(p). \tag{11.29}$$

Substituting Eq. (11.29) into Eq. (11.27), and after integrating over the three-momentum one finds

$$\Pi_{\mu\nu}(q^2) = (q_\mu q_\nu - q^2\,g_{\mu\nu})\frac{M_\rho^2}{f_\rho^2}\frac{1}{(q^2 - M_\rho^2 + \imath\epsilon)}. \tag{11.30}$$

The electromagnetic form factor of the pion, $F_\pi(q^2)|_{\text{VMD}}$, in VMD is given by

$$F_\pi(q^2)|_{\text{VMD}} = \frac{g_{\rho\pi\pi}}{f_\rho}\frac{M_\rho^2}{(M_\rho^2 - q^2)} = \frac{M_\rho^2}{(M_\rho^2 - q^2)} \tag{11.31}$$

with $F_\pi(0) = 1$, and $g_{\rho\pi\pi} \simeq f_\rho$ from data, so that $\Pi_{\mu\nu}(q^2)$ becomes

$$\Pi_{\mu\nu}(q^2) = -(q_\mu q_\nu - q^2\,g_{\mu\nu})\frac{1}{f_\rho^2}F_\pi(q^2). \tag{11.32}$$

The derivative of the pion form factor is related to the root-mean-squared radius so that

$$\frac{d}{dq^2} \Pi(s)_{uds}|_{s=0} = \frac{1}{6} \frac{1}{f_\rho^2} \langle r_\pi^2 \rangle = 0.076 \text{ GeV}^{-2}, \tag{11.33}$$

somewhat larger than the LQCD result, Eq. (11.23), and leading to

$$a_\mu^{HAD} = (775 \pm 14) \times 10^{-10}. \tag{11.34}$$

Subtracting a potential 5% contribution from the second derivative of the vector correlator (sign is negative for this term) would give

$$a_\mu^{HAD} = (736 \pm 14) \times 10^{-10}, \tag{11.35}$$

a value considerable larger than that obtained entirely from e^+e^- data [2]

$$a_\mu^{HAD}|_{e^+e^-} = (693.1 \pm 3.4) \times 10^{-10}. \tag{11.36}$$

A far better model for the pion form factor is the Dual-QCD$_\infty$ model involving an infinite number of zero width resonances, as in QCD$_\infty$, with masses and couplings given by the dual resonance model [11, 12]. The expression of the light-quark correlator is given by

$$\Pi_{uds}(s)|_{QCD_\infty} = \frac{1}{f_\rho^2} \frac{1}{\sqrt{\pi}} \frac{\Gamma(\beta - 1/2)}{\Gamma(\beta - 1)} B(\beta - 1, 1/2 - s/2M_\rho^2), \tag{11.37}$$

where β is a free parameter and $B(x, y)$ is the Euler beta-function. With $\beta = 2.30$ one obtains an excellent fit to the data up to $q^2 = -10\,\text{GeV}^2$, with a chi-squared per degree of freedom $\chi^2 = 1.5$ [11], and more importantly, a root-mean-square pion radius $< r_\pi^2 >= 0.436 \pm 0.004\,\text{fm}^2$ [11] to be compared with the most recent experimental value [13] $< r_\pi^2 >= 0.439 \pm 0.008\,\text{fm}^2$. The result for the anomaly in this model is

$$a_\mu^{HAD}|_{QCD_\infty} = (722 \pm 9) \times 10^{-10}, \tag{11.38}$$

in agreement with the LQCD prediction, Eq. (11.26), but not with the value from using e^+e^- data, Eq. (11.36). This result already incorporates the first and the second derivatives of the vector correlator.

References

1. F. Jegerlehner, Springer Tracts Mod. Phys. **274**, 1 (2017)
2. M. Davier, A. Hoecker, B. Malaescu, Z. Zhang, Eur. Phys. J. C **77**, 827 (2017)
3. S. Bodenstein, C.A. Dominguez, S.I. Eidelman, H. Spiesberger, K. Schilcher, J. High Energy Phys. **01**, 039 (2012)
4. S. Bodenstein, C.A. Dominguez, K. Schilcher, H. Spiesberger, Phys Rev. D **85**, 014029 (2012)
5. C.A. Dominguez, H. Horch, B. Jäger, N.F. Nasrallah, K. Schilcher, H. Spiesberger, H. Wittig, Phys Rev. D **96**, 074016 (2017)
6. S.J. Brodsky, E. de Rafael, Phys. Rev. **168**, 1620 (1968)
7. J.Z. Bai et al. BES Coll., Phys. Rev. Lett. **88**, 101802 (2002); M. Ablikim et al. BES Coll. Phys. Lett. B **677**, 239 (2009)
8. B. Chakraborty et al. Phys. Rev. D **89**, 114501 (2014); J. Koponen et al. Proc. of Sci. LATTICE-2014, 129 (2015)
9. B. Colquohoun et al., Phys. Rev. D **91**, 074514 (2015)
10. Sz. Borsanyi et al. Phys. Rev. D **96**, 074507 (2017)
11. C.A. Dominguez, Phys. Lett. B **512**, 331 (2001); C.A. Dominguez, T. Thapedi, J. High Energy Phys. **10**, 003 (2004); C.A. Dominguez, R. Röntsch, *ibid.* **10**, 085 (2007)
12. C. Bruch, A. Khodjamirian, J.H. Kühn, Eur. Phys. J. C **39**, 41 (2005)
13. S.R. Amendolia et al., Nucl. Phys. B **277**, 168 (1986)

Chapter 12
QCD Sum Rules at Finite Temperature

The extension of Finite Energy QCD sum rules (FESR) to finite temperature, and its many applications, has been reviewed recently in [1]. Hence, this section will be brief.

Introducing a finite temperature in quantum field theory implies the presence of a medium, the thermal bath, thus in principle breaking Lorentz covariance. To avoid this situation, and continue using covariant expressions for current correlators, one chooses a frame at rest with respect to the medium. This is to be understood in the sequel.

The starting point is the qualitative expectation on the behaviour of hadronic spectral functions with increasing temperature. Figure 12.1 shows a typical hadronic spectral function at $T = 0$ in the complex squared-energy, s-plane. It comprises a hadronically stable hadron (pole on the real s-axis), followed by a few resonances (poles on the second Riemann sheet), with hadronic widths increasing with energy. The pole gives a spectral function

$$\text{Im } \Pi(s)|_{POLE} = f_P^2 \, \delta(s - -M_P^2), \tag{12.1}$$

where f_P is the hadron-current coupling, and M_P its mass. The resonances, if narrow enough, could be parametrized e.g. by a Breit-Wigner form

$$\text{Im } \Pi(s)|_{RES} = f_R^2 \, \frac{M_R^3 \, \Gamma_R}{(s - M_R^2)^2 + M_R^2 \Gamma_R^2}, \tag{12.2}$$

where f_R is the resonance-current coupling entering a correlation function, M_R its mass, and Γ_R its (hadronic) width. At finite temperature this spectrum is expected to change dramatically with increasing T. An example is the pion pole with $f_P \equiv f_\pi$, which will acquire a width (interpreted as absorption in the thermal medium). The resonance widths should increase monotonically with temperature, and eventually become so large that there would be no trace of them in the spectrum. As this happens the hadron-current coupling should decrease with increasing T, approaching zero. Finally, the onset of perturbative QCD, i.e. the continuum, parametrized by s_0 (the radius of the integration contour in the complex s-plane) will decrease and approach the origin (Fig. 12.1). This scenario completes the qualitative description of quark-gluon deconfinement at finite temperature

© The Author(s), under exclusive licence to Springer Nature Switzerland AG 2018
C. A. Dominguez, *Quantum Chromodynamics Sum Rules*,
SpringerBriefs in Physics, https://doi.org/10.1007/978-3-319-97722-5_12

Fig. 12.1 Typical hadronic spectral function comprising a pole and three resonances. Curve **a** is at T=0, while curve **b** is at finite temperature

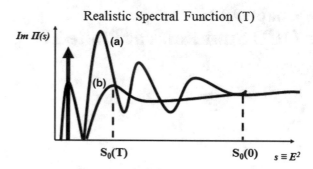

The starting point of the thermal QCD program is to allow for a temperature dependence in the quark propagator. There are two different procedures to achieve this, the Dolan-Jackiw [2], and the Matsubara formalism [3], equivalent at the one-loop level. The former leads to a straightforward extension of the QCD sum rule program through a simple correction to the fermion or boson propagator

$$S_F(k, T) = \frac{i}{k\!\!\!/ - m} - \frac{2\pi}{\left(e^{|k^0|/T} + 1\right)} (k\!\!\!/ + m)\, \delta(k^2 - m^2), \qquad (12.3)$$

and an equivalent expression for bosons, except for a positive relative sign between the two terms above, and the obvious replacement of the Fermi by the Bose thermal factor. An advantage of this expression is that it allows for a straightforward calculation of the imaginary part of current correlators, which is the function entering QCD sum rules. There is an issue at finite T, i.e. the potential presence of a space-like contribution to the imaginary part (absent at $T = 0$), to be discussed in Appendix E.

The extension of the QCD sum rule method to finite temperature was first proposed by Bochkarev and Shaposhnikov [4] in 1986, in the framework of Laplace transform QCD sum rules. One of the main problems with this kind of sum rules at finite T is that the role of s_0 is exponentially suppressed. While this might not have much impact at $T = 0$, it is definitely an undesired feature at $T \neq 0$. Since s_0 signals the end of the resonance region, and the threshold for QCD, and given the scenario shown in Fig. 12.1, its thermal behaviour should be enhanced rather than suppressed. For this reason thermal Finite Energy QCD sum rules (FESR) were proposed in [5, 6] and used over the years in a large number of applications (for a recent comprehensive review see [1]). A major breakthrough regarding the role of $s_0(T)$ took place recently with the establishment of a relation between this parameter and the Polyakov loop of Lattice QCD, which signals deconfinement [7].

A sample of results from thermal FESR is as follows. The T-dependence of the ratio $s_0(T)/s_0(0)$ from a FESR in the vector (ρ) channel, signalling deconfinement, is shown in Fig. 12.2 normalized to its $T = 0$ value (solid curve), together with the same ratio in the axial-vector channel (dash curve), signalling chiral symmetry restoration. The 10% difference at the end-point is within the accuracy of the method. The thermal

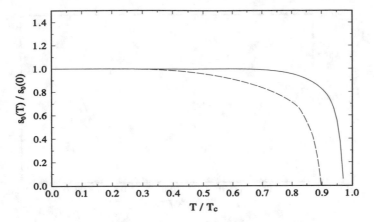

Fig. 12.2 The normalized thermal behaviour of $s_0(T)$ in the vector (ρ)-channel (solid curve), and in the axial-vector channel (dotted curve) [8]

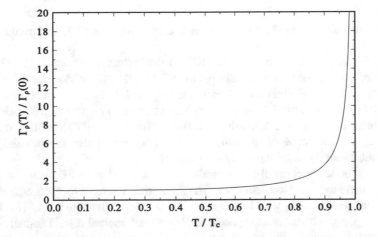

Fig. 12.3 The normalized thermal behaviour of the ρ-meson width [8]

behaviour of the normalized width of the ρ-meson [8] is shown in Fig. 12.3. The rise of the width near the deconfinement temperature is rather dramatic.

In the heavy-quark sector, i.e. charm- and bottom-quark region, results from thermal QCD sum rules turned out quite unexpected [9, 10]. Indeed, the thermal behaviour of $s_0(T)$, $\Gamma(T)$, and $f(T)$ clearly show that these states survive at the critical temperature for deconfinement. In fact, the deconfinement parameter, $s_0(T)$, begins to decrease near T_c, but eventually flattens out at $T/T_c \simeq 0.6$ up to $T/T_c \simeq 1.2$ beyond which there is no support to the integrals in the FESR. The width, while initially increasing substantially with temperature, it begins to decrease at and beyond T_c. Finally, the coupling remains constant up to close to T_c, and then shoots up. This scenario is further supported by results in the pseudoscalar (η_c) and scalar (χ_c)

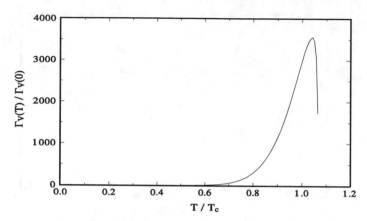

Fig. 12.4 Hadronic width of J/ψ as a function of T/T_c. The sum rules have no support beyond a certain temperature, normally for $T/T_c \lesssim 1$. Notice the exceptional behaviour for $T/T_c \gtrsim 1$

charmonium channels [11], as well as in vector bottonium (Υ), and pseudoscalar bottonium (η_c) [10].

Later determinations from Lattice QCD in the bottonium channel [12, 13] have confirmed this survival. These results put to rest three decades of popular belief in the opposite scenario, i.e. charmonium early melting [14]. In Fig. 12.4 the behaviour of the hadronic width of the J/ψ is shown as a function of T/T_c. The thermal behaviour of the other two parameters, the deconfinement parameter $s_0(T)$, and the coupling, $f_V(T)$, is also indicative of survival, i.e. $s_0(T)$ does not approach zero, and $f_V(T)$ grows substantially with increasing temperature.

The last issue concerns di-muon production in heavy-ion collisions in the energy region around the ρ-meson resonance. The theoretical issues of this process received some attention [18, 19] well before experiments were conducted [15, 16]. In this energy region the ρ-meson is expected to play a fundamental role. It actually enters the di-muon production rate through the pion electromagnetic form factor, and was considered in [19], albeit with hadronic parameters independent of temperature. The temperature dependence of the pion form factor was first proposed in [20], and rediscovered much later in [21, 22]. While the expression for the thermal width of the ρ-meson in [20] was somewhat crude, the conclusion was not, i.e. that there would be an important change in the di-muon production rate. More recently, given that the ρ-meson parameters at finite T had been obtained from FESR [1, 8], it became possible to obtain a parameter-free prediction of this rate. This is shown in Fig. 12.5. Given that the ρ-meson thermal parameters were already determined from the FESR in the light vector channel, the parameter-free prediction is exceptional.

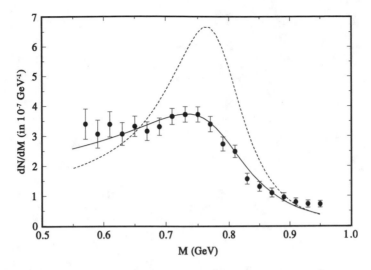

Fig. 12.5 The dimuon invariant mass distribution in In+In collisions in the region of the ρ-meson using pre-determined values of thermal parameters from QCDSR [1, 8] (solid curve). Dash curve is for all ρ-parameters independent of T. Data is from [15, 16]. Results are for $\mu = 0$. Finite chemical potential results change slightly in off-peak regions (see [17])

References

1. A. Ayala, C.A. Dominguez, M. Loewe, Adv. High Energy Phys. **2017**, 9291623 (2017)
2. L. Dolan, R. Jackiw, Phys. Rev. D **9**, 3320 (1974)
3. M. Le Bellac, *Thermal Field Theory*, (Cambridge University Press, United Kingdom, 1996)
4. A.I. Bochkarev, M.E. Shaposhnikov, Nucl. Phys. B **268**, 220 (1986)
5. C.A. Dominguez, M. Loewe, Phys. Lett. B **233**, 201 (1989)
6. A. Barducci, R. Casalbuoni, S. de Curtis, R. Gatto, G. Pettini, Phys. Lett. B **240**, 429 (1990)
7. J.P. Carlomagno, M. Loewe, Phys. Rev. D **95**, 036003 (2017)
8. A. Ayala, C.A. Dominguez, M. Loewe, Y. Zhang, Phys. Rev. D **86**, 114036 (2012)
9. C.A. Dominguez, M. Loewe, J.C. Rojas, Y. Zhang, Phys. Rev. D **81**, 014007 (2010)
10. C.A. Dominguez, M. Loewe, Y. Zhang, Phys. Rev. D **88**, 054015 (2013)
11. C.A. Dominguez, M. Loewe, J.C. Rojas, Y. Zhang, Phys. Rev. D **83**, 034033 (2011)
12. G. Aarts et al., J. High Ener. Phys. **03**, 084 (2013)
13. G. Aarts et al., J. High Ener. Phys. **12**, 064 (2013)
14. T. Matsui, H. Satz, Phys. Lett. B **178**, 416 (1986)
15. R. Arnaldi et al., NA60 collaboration. Phys. Rev. Lett. **96**, 162302 (2006)
16. S. Damjanovic, NA60 collaboration. Eur. Phys. J. **61**, 711 (2009)
17. A. Ayala, C.A. Dominguez, L.A. Hernandez, M. Loewe, A.J. Mizher, Phys. Rev. D **88**, 114028 (2013)
18. G. Domokos, J.I. Goldman, Phys. Rev. D **32**, 1109 (1985)
19. J. Cleymans, J. Fingberg, K. Redlich, Phys. Rev. D **35**, 2153 (1987)
20. C.A. Dominguez, M. Loewe, Z. Phys. C, Particles & Fields, **4**(9), 423 (1991)
21. V.L. Eletsky, B.L. Ioffe, J.I. Kapusta, Nucl. Phys. A **642**, 155 (1998)
22. V.L. Eletsky, B.L. Ioffe, J.I. Kapusta, Eur. Phys. J. A **3**, 381 (1998)

Chapter 13
Summary and Outlook

In this book the current method of QCD sum rules, Finite Energy Sum Rules (FESR), has been introduced and reviewed. This approach is based on the relation between QCD and hadronic physics, established by invoking Cauchys residue theorem in the complex squared-energy s-plane [1]. In this plane QCD information is present on the circular contour of radius $s = |s_0|$ (see Fig. 1.1, Sect. 1), while hadronic physics is formulated along the positive real semi-axis. The objects considered in this plane are current correlators, e.g. Eq. (1.5), having both a QCD as well as a hadronic representation in terms of their respective quantum fields. This method complements Lattice QCD, achieving comparable precision in many important instances.

High precision applications are mostly restricted to two-point functions, as they are currently known up to $\mathcal{O}(\alpha_f^\triangle)$ in perturbative QCD. In contrast, there is no information available on radiative corrections to three-point functions. Even at next-to-leading order, $\mathcal{O}(\alpha_s)$, these corrections would lead to considerable improvement of current predictions of form factors. For instance, the electromagnetic form factors of the proton, $F_{1,2}(q^2)$, determined from a three-point function sum rule, shows only the electric form factor, $F_1(q^2)$, to agree well with data [2]. In contrast, the form factor $F_2(q^2)$ disagrees with data by a factor two. The reason could well be the absence of radiative corrections. Nevertheless, it would be important to understand why three-point function sum rules for mesons, instead of baryons, do lead to reasonable predictions, e.g. for the pion electromagnetic form factor [3, 4], the axial-vector coupling of the nucleon [5], the $\rho - \pi\pi$ strong interaction coupling [4], and the strong $\omega\rho\pi$ coupling [6].

A topic of current interest, omitted here due to space limitations, is that of the leptonic decay constants of heavy pseudoscalar mesons, such as the D-, D_s-, B-, B_s- and B_c-mesons. A pioneering determination using Hilbert moment QCD sum rules [7] was followed by many determinations over the years (for recent results see [8, 9] and references therein). The leptonic decay constant of the B_c meson remains a contentious issue, as results from QCD sum rules differ substantially among each other.

Current state of the art determinations from FESR include the values of the light and heavy quark masses (except the top-quark). Improvement in precision could be achieved by employing series convergence methods. For instance, it has been shown

© The Author(s), under exclusive licence to Springer Nature Switzerland AG 2018
C. A. Dominguez, *Quantum Chromodynamics Sum Rules*,
SpringerBriefs in Physics, https://doi.org/10.1007/978-3-319-97722-5_13

that Padè approximants do improve convergence of the perturbative series [10]. However, there is more than this particular choice in the literature to be explored.

A topic of current high importance is that of the leading hadronic contribution to the anomalous magnetic moment of the muon, $g - 2$, as discussed in Chap. 11. The QCD sum rule approach allows for an entirely theoretical prediction of this quantity, provided the first two derivatives of the vector current correlator at the origin are known. It is important to stress that these derivatives can be obtained analytically entirely from QCD for the charm- and bottom-quark regions. There is perfect agreement between the QCD sum rule value of the anomaly in these regions and LQCD results. A model independent determination in the light-quark sector is only possible in the framework of Lattice QCD. Current precision in this framework [11] is not enough to draw definite conclusions regarding the need for Physics beyond the Standard Model (SM). It is only through relations between the vector current correlator and the pion electromagnetic form factor that these derivatives can be estimated. Results from this approach at the moment are consistent with the SM, but disagree with results from using data on electron-positron annihilation into hadrons. Potential problems with this data base have been pointed out in Sect. 8, based on the analysis of [12].

An important extension of the QCD sum rule method is to consider finite temperature, as first proposed in [13], and briefly discussed in Sect. 12. A recent comprehensive review [14] should be consulted. Perhaps the most successful and unexpected result was the prediction of the survival of charmonium and bottonium states at, and beyond the critical temperature for deconfinement [15, 16]. This prediction was later confirmed by Lattice QCD results in the bottom-quark sector [17, 18]. Such a scenario contradicts and lays to rest a decades long popular expectation of charmonium melting at or even below the critical temperature [19]. Occam's razor applies.

It is also possible to formulate QCD sum rules at finite temperature as well as baryon chemical potential [20]. This allows for an exploration of the QCD phase diagram and the study of the phase transitions of chiral symmetry restoration and deconfinement. Regarding the latter it is important to highlight a recent result [21] showing a relation between the QCD sum rule deconfinement parameter $s_0(T)$ (the onset of perturbative QCD) and the Polyakov loop, i.e. the LQCD deconfinement parameter. This unexpected relation provides a fundamental validation of the thermal QCD sun rule method.

The current research direction of finite temperature QCD is to combine it with the presence of external strong magnetic fields. At present there is a substantial amount of information in the framework of hadronic models, e.g. Linear Sigma Model, Nambu-Jona-Lasinio Model, etc. (for reviews see e.g. [22, 23]). On the theory side the first QCD sum rule attempt at including a magnetic field was done at zero temperature [24]. The addition of temperature dependence is currently work in progress.

For background material on QCD sum rules the reader may consult the textbooks [25, 26].

References

1. R. Shankar, Phys. Rev. D **15**, 755 (1977)
2. H. Castillo, C.A. Dominguez, M. Loewe, J. High Ener, Phys. **0503**, 012 (2005)
3. B.L. Ioffe, A.V. Smilga, Nucl. Phys. B **216**, 373 (1983)
4. C.A. Dominguez, M.S. Fetea, M. Loewe, Phys. Lett. B **406**, 149 (1997)
5. C.A. Dominguez, M. Loewe, C. van Gend, Phys. Lett. B **460**, 442 (1999)
6. C.A. Dominguez, M. Loewe, Phys. Lett. B **481**, 295 (2000)
7. C.A. Dominguez, N. Paver, Phys. Lett. B **197**, 423 (1987)
8. M.J. Baker, J. Bordes, C.A. Dominguez, J. Peñarrocha, K. Schilcher, J. High Ener. Phys. **07**, 032 (2014)
9. W. Lucha, D. Melikhov, S. Simula, Phys. Rev. D **88**, 056011 (2013)
10. S. Bodenstein, C.A. Dominguez, K. Schilcher, J. High Ener. Phys. **07**, 138 (2013)
11. C.A. Dominguez, H. Horch, B. Jäger, N.F. Nasrallah, K. Schilcher, H. Spiesberger, H. Wittig, Phys Rev. D **96**, 074016 (2017)
12. S. Bodenstein, C.A. Dominguez, S.I. Eidelman, H. Spiesberger, K. Schilcher, J. High Energy Phys. **01**, 039 (2012)
13. A.I. Bochkarev, M.E. Shaposhnikov, Nucl. Phys. B **268**, 220 (1986)
14. A. Ayala, C.A. Dominguez, M. Loewe, Adv. High Energy Phys. **2017**, 9291623 (2017)
15. C.A. Dominguez, M. Loewe, J.C. Rojas, Y. Zhang, Phys. Rev. D **81**, 014007 (2010)
16. C.A. Dominguez, M. Loewe, Y. Zhang, Phys. Rev. D **88**, 054015 (2013)
17. G. Aarts et al., J. High Ener. Phys. **03**, 084 (2013)
18. G. Aarts et al., J. High Ener. Phys. **12**, 064 (2013)
19. T. Matsui, H. Satz, Phys. Lett. B **178**, 416 (1986)
20. A. Ayala, A. Bashir, C.A. Dominguez, E. Gutierrez, M. Loewe, A. Raya, Phys. Rev. D **84**, 056004 (2011)
21. J.P. Carlomagno, M. Loewe, Phys. Rev. D **95**, 036003 (2017)
22. V.A. Miranski, I.A. Shovkovy, Phys. Rep. **576**, 1 (2015)
23. J.O. Andersen, W.R. Naylor, A. Tranberg, Rev. Mod. Phys. **88**, 025001 (2016)
24. A. Ayala, C.A. Dominguez, L.A. Hernandez, M. Loewe, J.C. Rojas, C. Villavicencio, Phys. Rev. D **92**, 016006 (2015)
25. P. Pascual, R. Tarrach, *QCD: Renormalization for the Practitioner*, Springer-Verlag, Brlin Heidelberg (1984)
26. W. Greiner, A. Schäfer, *Quantum Chromodynamics* (Springer-Verlag, New York, Brlin, Heidelberg, 1995)

Appendix A
Dressed Propagators and Selected Integrals

The quark propagator in an external non-Abelian gauge field is given by [1]

$$
S_F^{\alpha\beta}(p,m) = \frac{\delta^{\alpha\beta}}{\not{p} - m} - \frac{g}{2}\left(\frac{\lambda^a}{2}\right)^{\alpha\beta} \frac{G_{\kappa\lambda}^a}{(p^2-m^2)^2}\left[p_\phi \epsilon^{\phi\kappa\lambda\tau}\gamma_\tau\gamma_5 + m\sigma^{\kappa\lambda}\right]
$$
$$
+ \frac{g^2}{12}\delta^{\alpha\beta}\, G^2\, m\, \frac{p^2 + m\not{p}}{(p^2-m^2)^4}, \tag{A.1}
$$

where m is the quark mass, $\alpha,\beta = 1, 2, N_c$ are colour indexes, $a = 1,2,3...8$ the $SU(3)_c$ index, g is the strong quark-gluon coupling, and $G_{\mu\nu}^a$ the gluon field tensor. In the massless limit this becomes

$$
S_F^{\alpha\beta}(p,m)|_{m\to0} = \frac{\not{p}+m}{p^2-m^2}\delta^{\alpha\beta} - \frac{g}{2}\left(\frac{\lambda^a}{2}\right)^{\alpha\beta} G_{\kappa\lambda}^a \frac{1}{p^4}\, p_\phi\, \epsilon^{\phi\kappa\lambda\tau}\,\gamma_\tau\,\gamma_5. \tag{A.2}
$$

In coordinate space, and in the massless limit, the quark propagator is given by

$$
S_F^{\alpha\beta}(x,m)|_{m\to0} = \frac{\delta^{\alpha\beta}}{2\pi^2}\left[\frac{\gamma^\mu \cdot x_\mu}{x^4} + \frac{i}{2}\frac{m}{x^2}\right] - \frac{1}{8\pi^2}\frac{x^\phi}{x^2}\frac{g}{2}\left(\frac{\lambda^a}{2}\right)^{\alpha\beta} G_{\kappa\lambda}^a \epsilon^{\phi\kappa\lambda\tau}\gamma_\tau\,\gamma_5. \tag{A.3}
$$

The quark condensate in an external non-Abelian gauge filed is

$$
\langle 0|\bar{q}_i^\alpha(x)\, q_j^\beta(0)|0\rangle = \frac{1}{12}\delta^{\alpha\beta}\left[\delta_{ij} + \frac{i}{4}m_q\, x^\mu\,(\gamma_\mu)_{ji} + \mathcal{O}(m_q^2)\right]\langle 0|\bar{q}\, q|0\rangle
$$
$$
+ \frac{\delta^{\alpha\beta}}{12}\left(-\frac{i}{16}\right) x^2\left(\delta_{ij} + \frac{i}{6}m_q\, x^\mu\,(\gamma_\mu)_{ji}\right)\left\langle 0|\bar{q}\,\sigma^{\mu\nu}\, i\, g\frac{\lambda^a}{2}\, G_{\mu\nu}^a\, q|0\right\rangle, \tag{A.4}
$$

where $\langle 0|\bar{q}\, q|0\rangle$ is the standard quark condensate, the second term in the expansion is the dimension $d = 5$ mixed quark-gluon condensate, and the dimension $d = 6$ four-quark condensate has been omitted. The suggestion of relating the four-quark condensate to the square of the ordinary quark condensate (vacuum saturation) [2] has no theoretical support [3]. Furthermore, the numerical value of this dimension

© The Author(s), under exclusive licence to Springer Nature Switzerland AG 2018
C. A. Dominguez, *Quantum Chromodynamics Sum Rules*,
SpringerBriefs in Physics, https://doi.org/10.1007/978-3-319-97722-5

$d = 6$ term in the OPE, as determined from FESR in the vector channel, has some 100% uncertainty [4]. The situation is better for the $d = 6$ chiral condensate (V-A), known with some 10% uncertainty [5], albeit an order of magnitude smaller than the quark condensate at relevant comparable energy scales.

The following space-time integrals enter in current correlators:

$$\int d^4x \, e^{iq \cdot x} \frac{1}{x^2} = -\frac{4\pi^2 i}{q^2} \tag{A.5}$$

$$\int d^4x \, e^{iq \cdot x} \frac{x^\mu}{x^2} = \frac{8\pi^2}{(q^2)^2} q^\mu \tag{A.6}$$

$$\int d^4x \, e^{iq \cdot x} \frac{x^\mu x^\nu}{x^2} = \frac{8\pi^2 i}{(q^2)^3} (-q^2 g^{\mu\nu} + 4 q^\mu q^\nu) \tag{A.7}$$

$$\int d^4x \, e^{iq \cdot x} \frac{x^\mu}{x^4} = \frac{2\pi^2}{q^2} q^\mu \tag{A.8}$$

$$\int d^4x \, e^{iq \cdot x} \frac{x^\mu x^\nu}{x^4} = \frac{2\pi^2 i}{(q^2)^2} (-q^2 g^{\mu\nu} + 2 q^\mu q^\nu) \tag{A.9}$$

$$\int d^4x \, e^{iq \cdot x} \frac{1}{(x^2)^n} = i \frac{\pi^2}{2^{2n-4}} \frac{(-)^n}{\Gamma(n)} \frac{1}{\Gamma(n-1)} \frac{\ln(-q^2/\mu^2)}{(q^2)^{2-n}} \quad (n > 1) \tag{A.10}$$

$$\int d^4x \, e^{iq \cdot x} \frac{x^\mu}{(x^2)^n} = \frac{\pi^2}{2^{2n-5}} \frac{(-)^n}{\Gamma(n)} q^\mu \frac{(n-2)}{\Gamma(n-1)} \frac{\ln(-q^2/\mu^2)}{(q^2)^{3-n}} \quad (n > 2) \tag{A.11}$$

$$\int d^4x \, e^{iq \cdot x} \frac{x^\mu x^\nu}{(x^2)^n} = i \frac{\pi^2}{2^{2n-5}} \frac{(-)^{n+1}}{\Gamma(n)} \frac{(n-2)}{\Gamma(n-1)}$$
$$\times \left(g^{\mu\nu} + 2(n-3) \frac{q^\mu q^\nu}{q^2} \right) \frac{\ln(-q^2/\mu^2)}{(q^2)^{3-n}} \quad (n > 2) \tag{A.12}$$

$$\int d^4x \, e^{iq \cdot x} \frac{x^\alpha x^\beta x^\rho}{(x^2)^n} = \frac{\pi^2}{2^{2n-6}} (-)^{n+1} \frac{(n-2)(n-3)}{\Gamma(n)\Gamma(n-1)} (q^2)^{n-4}$$
$$\times \left(q^\alpha g^{\beta\rho} + q^\beta g^{\alpha\rho} + q^\rho g^{\alpha\beta} + 2 \frac{(n-4)}{q^2} q^\alpha q^\beta q^\rho \right) \ln(-q^2/\mu^2). \tag{A.13}$$

Dimensional regularized integrals, with dimension $D = 4 + 2\epsilon$:

$$\int d^D k \, (k^2)^a = 0, \tag{A.14}$$

valid for all a, except $a = D/2$. Hence, for consistency one defines

$$\int d^D k \, (k^2)^{-D/2} = 0. \tag{A.15}$$

$$\int d^D k \, \frac{(k^2)^a}{(k^2 + M^2)^b} = \pi^{D/2} \, (M^2)^{(a-b+D/2)} \, \frac{\Gamma(a + D/2) \, \Gamma(b - a - D/2)}{\Gamma(D/2) \, \Gamma(b)} \tag{A.16}$$

$$\int \frac{d^4 k}{(2\pi)^4} \, \frac{1}{k^2 \, (k - q)^2} = \frac{-i}{(4\pi)^2} \left[\ln(-q^2/\mu^2) - 2 \right] \tag{A.17}$$

$$\int \frac{d^4 k}{(2\pi)^4} \, \frac{k^\mu}{k^2 \, (k - q)^2} = \frac{-i}{(4\pi)^2} \, \frac{1}{2} \, q^\mu \left[\ln(-q^2/\mu^2) - 2) \right] \tag{A.18}$$

$$\int \frac{d^4 k}{(2\pi)^4} \, \frac{k^\mu \, k^\nu}{k^2 \, (k - q)^2} = \frac{i}{(4\pi)^2} \left[q^2 \, g^{\mu\nu} \left(\frac{\ln(-q^2/\mu^2)}{12} - \frac{2}{9} \right) \right.$$
$$\left. + q^\mu q^\nu \left(-\frac{\ln(-q^2/\mu^2)}{3} + \frac{13}{18} \right) \right] \tag{A.19}$$

$$\int \frac{d^4 k}{(2\pi)^4} \, \frac{1}{k^2 \, [(k - q)^2 - m^2]} = \frac{-i}{(4\pi)^2} \left[\ln \left(\frac{-q^2}{\mu^2} \right) + \frac{m^2}{q^2} \, \ln \left(\frac{m^2}{-q^2} \right) \right.$$
$$\left. + \left(1 - \frac{m^2}{q^2} \right) \ln \left(1 - \frac{m^2}{q^2} \right) - 2 \right] \tag{A.20}$$

$$\int \frac{d^4 k}{(2\pi)^4} \, \frac{k^\mu}{k^2 \, [(k - q)^2 - m^2]} = \frac{-i}{(4\pi)^2} \, \frac{1}{2} \, q^\mu \left[\ln \left(\frac{-q^2}{\mu^2} \right) + \frac{m^2}{q^2} \left(2 - \frac{m^2}{q^2} \right) \right.$$
$$\times \ln \left(\frac{m^2}{-q^2} \right) + \left(1 - 2 \frac{m^2}{q^2} + \frac{m^4}{q^4} \right) \ln \left(1 - \frac{m^2}{q^2} \right) + \frac{m^2}{q^2} - 2 \right]. \tag{A.21}$$

$$I(q, a, b) \equiv \frac{1}{\mu^{2\epsilon}} \int \frac{d^D p}{(2\pi)^D} \, \frac{1}{(p^2 + i\eta)^a \, [(p - q)^2 + i\eta]^b}$$
$$= \frac{i}{(4\pi)^2} \left(-\frac{q^2}{4\pi\mu^2} \right)^\epsilon \, q^{-2(a+b-2)} \, \frac{\Gamma(2 - a + \epsilon) \, \Gamma(2 - b + \epsilon)}{\Gamma(a)\Gamma(b)\Gamma(4 - a - b + 2\epsilon)}$$
$$\times \, \Gamma(a + b - 2 - \epsilon). \tag{A.22}$$

$$I^\mu(q, a, b) \equiv \frac{1}{\mu^{2\epsilon}} \int \frac{d^D p}{(2\pi)^D} \, \frac{p^\mu}{(p^2 + i\eta)^a \, [(p - q)^2 + i\eta]^b}$$
$$= \frac{i}{(4\pi)^2} \left(-\frac{q^2}{4\pi\mu^2} \right)^\epsilon \, q^{-2(a+b-2)} \, q^\mu \, \frac{\Gamma(3 - a + \epsilon) \, \Gamma(2 - b + \epsilon)}{\Gamma(a)\Gamma(b)\Gamma(5 - a - b + 2\epsilon)}$$
$$\times \, \Gamma(a + b - 2 - \epsilon). \tag{A.23}$$

$$I^{\mu\nu}(q, a, b) \equiv \frac{1}{\mu^{2\epsilon}} \int \frac{d^D p}{(2\pi)^D} \frac{p^\mu p^\nu}{(p^2 + i\eta)^a [(p-q)^2 + i\eta]^b}$$

$$= \frac{i}{(4\pi)^2} \left(-\frac{q^2}{4\pi\mu^2} \right)^\epsilon q^{-2(a+b-2)} \left[g^{\mu\nu} q^2 \frac{\Gamma(3-a+\epsilon)\,\Gamma(3-b+\epsilon)}{2\,\Gamma(a)\,\Gamma(b)\,\Gamma(6-a-b+2\epsilon)} \right.$$

$$\times\, \Gamma(a+b-3+\epsilon) + q^\mu q^\nu \frac{\Gamma(4-a+\epsilon)\,\Gamma(2-b+\epsilon)}{\Gamma(a)\,\Gamma(b)\,\Gamma(6-a-b+2\epsilon)}$$

$$\left. \times\, \Gamma(a+b-2+\epsilon) \right]. \tag{A.24}$$

In some applications of QCDSR one uses the imaginary part of current correlators, which involve integrals of delta-functions. A sample list is provided below (for more details see [6]).

Definitions

$$I_1 \equiv \int dp_0 \; \theta(p_0) \; \delta(p^2 - m^2) = \frac{1}{2E}, \qquad E = +\sqrt{(\mathbf{p}^2 + m^2)}. \tag{A.25}$$

$$w(a, b, c) \equiv [a^2 + b^2 + c^2 - 2(ab + ac + bc)]^{1/2}$$
$$= [(a + b - c)^2 - 4ab]^{1/2}$$
$$= [a - (\sqrt{b} + \sqrt{c})^2]^{1/2} \, [a - (\sqrt{b} - \sqrt{c})^2]^{1/2}. \tag{A.26}$$

Integrals

$$I_2 \equiv \int d^4q \; \theta(p^0 - q^0) \; \delta[(p-q)^2 - m^2] \; \delta(q^2 - M^2)$$

$$= \frac{\pi}{2} \theta(p^0) \; \theta[p^2 - (M + m)^2] \frac{1}{p^2} \; w(p^2, M^2, m^2). \tag{A.27}$$

$$I_3 \equiv \int d^4q \; \theta(q^0) \; \theta(p^0 - q^0) \; \delta[(p-q)^2 - m^2] \; \delta(q^2 - M^2) \; q^\mu$$

$$= \frac{\pi}{4} \theta(p^0) \; \theta[p^2 - (M + m)^2] \frac{1}{p^4} \; w(p^2, M^2, m^2) \; p^\mu$$

$$\times\, (p^2 + M^2 - m^2) \tag{A.28}$$

$$I_4 \equiv \int d^4q \; \theta(q^0) \; \theta(p^0 - q^0) \; \delta[(p-q)^2 - m^2] \; \delta(q^2 - M^2) \; q^\mu q^\nu$$

$$= \frac{\pi}{6} \theta(p^0) \; \theta[p^2 - (M + m)^2] \frac{1}{p^4} \; w(p^2, M^2, m^2) \times \left[\frac{p^\mu p^\nu}{p^2} \right.$$

$$\left. \times\, [(p^2 + M^2 - m^2)^2 - p^2 M^2] - \frac{g^{\mu\nu}}{4} \, [w(p^2, M^2, m^2)]^2 \right]. \tag{A.29}$$

Appendix B
Current Correlators in QCD

In this Appendix it is shown how to obtain a current correlator to leading order in perturbative QCD, as well as to leading order in the power corrections due to the quark condensate and the gluon condensate.

Starting with the light-quark, electrically charged, vector-current correlator in the chiral limit ($m_{u,d} = 0$), identical to the axial-vector current correlator in this limit, it is defined as

$$\Pi_{\mu\nu}^{V}(q^2) = i \int d^4x \, e^{iq \cdot x} \, \langle 0 | T(J_\mu(x) J_\nu^\dagger(0)) | 0 \rangle \,, \tag{B.1}$$

where

$$J_\mu(x) =: \bar{d}_i^a(x) \, \gamma_\mu|_{ij} \, u_j^a(x), \tag{B.2}$$

The time-order product in Eq. (B.1) is given by

$$\tau \equiv T(J_\mu(x) J_\nu^\dagger(0)) = \bar{d}_i^a(x) \, \gamma_\mu|_{ij} \, u_j^a(x) \, \bar{u}_k^b(0) \, \gamma_\nu|_{kl} \, d_l^b(0). \tag{B.3}$$

Contracting the quark fields generates the quark propagators, $S_F(x)$ so that τ becomes

$$\tau = (-1)i \, S_F^{(d)}(-x)|_{li} \, \gamma_\mu|_{ij} \, \delta^{ab} \, i \, S_F(x)|_{jk} \, \delta^{ab} \, \gamma_\nu|_{kl}, \tag{B.4}$$

where the Wick-sign, (-1), is due to the ordering of the quark fields. After contracting $\delta^{ab} \delta ab = N_c$ (the number of colours) the trace becomes

$$\tau = N_c \, Tr[S_F^{(d)}(-x) \gamma_\mu S_F(x) \gamma_\nu]. \tag{B.5}$$

Invoking Eq. (A.3) for the quark propagator in the massless limit, the correlator becomes

$$\Pi_{\mu\nu}^{V}(q^2) = N_c \, Tr(\gamma_\mu \, \gamma_\alpha \, \gamma_\nu \, \gamma_\beta) \,] \int \frac{d^4x}{4\,\pi^4} \, e^{iq \cdot x} \, \frac{x^\alpha \, x^\beta}{x^8}. \tag{B.6}$$

© The Author(s), under exclusive licence to Springer Nature Switzerland AG 2018
C. A. Dominguez, *Quantum Chromodynamics Sum Rules*,
SpringerBriefs in Physics, https://doi.org/10.1007/978-3-319-97722-5

Using Eq. (A.12) with $n = 4$ and performing the trace gives

$$\Pi_{\mu\nu}^{V}(q^2) = -\frac{1}{4\pi^2}\ln(-q^2/\mu^2)\,(q_\mu q_\nu - q^2\,g_{\mu\nu}). \tag{B.7}$$

Often in the literature the electrically neutral vector current is considered

$$V_\mu(x) = \frac{1}{2} : [\bar{u}(x)\,\gamma_\mu\,u(x) - \bar{d}(x)\,\gamma_\mu\,d(x)]. \tag{B.8}$$

This choice leads to an extra factor of $\frac{1}{2}$ in Eq. (B.7), and is the source of much confusion.

Next, the light-quark pseudoscalar current correlator $\psi_5(q^2)$, Eq. (1.5), involves the time-ordered product

$$\tau \equiv T[\partial^\mu A_\mu(x)\,\partial^\nu A_\nu^\dagger(0)] = -i^2 \bar{d}_i^a(x)(\gamma_5)_{ij}\,u_j^a(x)\,\bar{u}_k^b(0)(\gamma_5)_{kl}\,d_l^b(0), \tag{B.9}$$

where the overall minus sign is due to the Wick-sign. This trace then becomes

$$\tau = -i^2\,N_c\,Tr\left[S_F^{(d)}(-x)\,\gamma_5\,S_F^{(u)}(x)\,\gamma_5\right]. \tag{B.10}$$

In momentum-space this is given by

$$\tau = -i^2\,N_c\,4\,(m_u m_d - k_1 \cdot k_2), \tag{B.11}$$

where the term of order $\mathcal{O}(m_q^2)$ can be safely neglected (fully justified after comparing this perturbative results with the power corrections due to the quark and the gluon condensates).

The correlator $\psi_5(q^2)$ is

$$\psi_5(q^2) = 4\,i^3\,N_c\,(m_u + m_d)^2 \int d^4x\,e^{iq\cdot x} \int \frac{d^4k_1}{(2\pi)^4} \int \frac{d^4k_2}{(2\pi)^4}\,\frac{(k_1 \cdot k_2)}{k_1^2\,k_2^2}\,e^{i(k_2-k_1)\cdot x}. \tag{B.12}$$

After performing the first (space-time) integral, and making use of the emerging delta-function $\delta(q + k_2 - k_1)$ to integrate over e.g. k_2 one has

$$\psi_5(q^2) = -4\,i\,N_c\,(m_u + m_d)^2 \int \frac{d^4k}{(2\pi)^4}\,\frac{(k^2 - k\cdot q)}{k^2(k - q)^2}. \tag{B.13}$$

The first integral above vanishes, while the second integral is given in Eq. A.18 so that

$$\psi_5(q^2) = (m_u + m_d)^2 \left(\frac{3}{8\pi^2}\right)(-q^2)\,\ln(-q^2/\mu^2). \tag{B.14}$$

Current information on the PQCD expression of $\psi_5(q^2)$ beyond the leading order is given in Appendix C.

Next, the quark condensate contribution to $\psi_5(q^2)$ is obtained as follows. Starting with the time ordered product, Eq. (B.17), for each flavour a quark-anti-quark pair combines to make a quark propagator while the other pair interacts with the QCD vacuum giving rise to a quark condensate, $\langle q_j^a(x)\, \bar{q}_k^b(0)\rangle$, giving

$$\tau = i^3\, \delta^{ab}\, (\gamma_5))_{ij} (\gamma_5))_{kl}\Big[S_F^d(-x)|_{li} \langle u_j^a(x)\bar{u}_k^b(0)\rangle + S_F^u(x)|_{jk}\langle \bar{d}_i^a(x)d_l^b(0)\rangle\Big]. \quad (B.15)$$

From Eq. (A.3) the quark propagator is

$$S_F(x) = \frac{1}{2\,\pi^2}\frac{\gamma^\alpha\, x_\alpha}{x^4} + i\,\frac{1}{4\pi^2}\frac{m_q}{x^2} \quad (B.16)$$

which after substitution in the trace leads to

$$\tau = i\, Tr\Big[S_F^d(-x)\,\gamma_5\, \langle \bar{u}^a(0)u^a(x)\rangle\,\gamma_5 + S_F^u(x)\,\gamma_5\,\langle \bar{d}^a(x)\,d^a(0)\,\gamma_5\rangle\Big], \quad (B.17)$$

where

$$S_F^q(x) = \frac{1}{2\,\pi^2}\frac{\gamma_\mu\, x^\mu}{x^4} + i\frac{m_q}{4\,\pi^2}\frac{1}{x^2}, \quad (B.18)$$

and

$$\langle \bar{q}^a q^a\rangle = \frac{1}{12}\Big(1 + \frac{i}{4}\,m_q\,\gamma^\mu\Big). \quad (B.19)$$

After substituting the above two expressions in the trace, Eq. (B.17), and subsequently substituting the trace in Eq. (1.5), and performing the coordinate space integral the result for $\psi_5(q^2)$ is

$$\psi_5(q^2)|_{\langle \bar{q}q\rangle} = \frac{(m_u + m_d)^2}{2\,q^2}\big[m_d\langle \bar{u}u\rangle + m_u\langle \bar{d}d\rangle\big]. \quad (B.20)$$

In practical applications $\langle \bar{u}u\rangle = \langle \bar{d}d\rangle$ is a very good approximation. In fact, SU(2) vacuum symmetry breaking is negligible in comparison with explicit symmetry breaking $(m_d/m_u \simeq 2)$.

The quark condensate contribution to the vector current correlator, Eq. (B.1), is obtained in a similar fashion, with the result

$$\Pi_{\mu\nu}^V(q^2)|_{\langle \bar{q}q\rangle} = \frac{1}{q^4}\,(m_u + m_d)\langle \bar{q}\, q\rangle\, (q_\mu\, q_\nu - q^2\, g_{\mu\nu}), \quad (B.21)$$

where vacuum symmetry, $\langle \bar{u}u\rangle \simeq \langle \bar{d}d\rangle$, has been used.

The next power correction in the OPE is the gluon condensate. Let us consider the correlation function in the light-quark vector channel, Eq. (B.1). Contracting it with $g^{\mu\nu}$ projects the scalar function $\Pi_V(q^2)$

$$g^{\mu\nu} \, \Pi^V_{\mu\nu}(q^2) = -3 \, q^2 \, \Pi_V(q^2), \tag{B.22}$$

where $\Pi_V(q^2)$ is explicitly given by

$$\Pi_V(q^2) = (-1)^2 \, \frac{i^3}{3 \, q^2} \, (\gamma_\mu)_{ij} \, (\gamma^\mu)_{kl} \int d^4x \, e^{iq \cdot x} \int \frac{d^4k_1}{(2\pi)^4} \, e^{-ik_1 \cdot x}$$
$$\times \int \frac{d^4k_2}{(2\pi)^4} \, e^{ik_2 \cdot x} \, S^{ab}_{jk}(k_1) \, S^{ba}_{li}(k_2), \tag{B.23}$$

where the additional factor i arises from the Wick sign. After integrating over space-time $\Pi_V(q^2)$ becomes

$$\Pi_V(q^2) = \frac{i^3}{3 \, q^2} \, (\gamma_\mu)_{ij} \, (\gamma^\mu)_{kl} \, (2\pi)^4 \int \frac{d^4k_1}{(2\pi)^4} \, e^{-ik_1 \cdot x} \int \frac{d^4k_2}{(2\pi)^4} \, e^{ik_2 \cdot x}$$
$$\times S^{\alpha\beta}_{jk}(k_1) \, S^{\beta\alpha}_{li}(k_2) \, \delta^{(4)}(q + k_2 - k_1). \tag{B.24}$$

The dressed quark propagator, Eq. (8.18), in the chiral limit and only up to order $\mathcal{O}(G^a_{\kappa\lambda})$ is given by

$$S^{\alpha\beta}_{jk}(k) = \frac{\not{k}_{jk}}{k^2} \, \delta^{\alpha\beta} - \frac{g}{2} \left(\frac{\lambda^a}{2} \right)^{\alpha\beta} \frac{G^a_{\kappa\lambda}}{k^4} \, k^\phi \, \epsilon^{\phi\kappa\lambda\tau} (\gamma_\tau \, \gamma_5)_{jk}. \tag{B.25}$$

After substituting this quark propagator into Eq. (B.24) gives

$$\Pi_V(q^2)|_{\langle G^2 \rangle} = \frac{i^3}{3 \, q^2} \, (2\pi)^4 (\gamma_\mu)_{ij} \, (\gamma^\mu)_{kl} \left(\frac{g}{2} \right)^2 \left(\frac{\lambda^a}{2} \right) \left(\frac{\lambda^b}{2} \right) \left\langle G^a_{\kappa\lambda} \, G^b_{\rho\delta} \right\rangle \epsilon^{\phi\kappa\lambda\tau}$$
$$\times \epsilon^{\nu\rho\delta\tau} (\gamma_\tau \gamma_5)_{jk} (\gamma_\alpha \gamma_5)_{li} \int \frac{d^4k_1}{(2\pi)^4} \int \frac{d^4k_2}{(2\pi)^4} \, \frac{k_1^\phi \, k_2^\nu}{k_1^4 \, k_2^4} \, \delta^{(4)}(q + k_2 - k_1). \tag{B.26}$$

The gluon condensate term becomes proportional to $\langle 0 | G^2 | 0 \rangle > \equiv \langle G^2 \rangle$ after using the relation

$$\left\langle G^a_{\kappa\lambda} \, G^b_{\rho\delta} \right\rangle = \frac{\delta^{ab}}{96} \left(g_{\kappa\rho} \, g_{\lambda\delta} - g_{\kappa\delta} \, g_{\lambda\rho} \right) \langle G^2 \rangle. \tag{B.27}$$

The trace of the six gamma-matrices gives a factor $-8 \, g_{\tau\alpha}$, so that the index α in the epsilon tensor becomes τ, and the contraction of the two epsilon tensors with the metric term in parenthesis above reduce to

$$\epsilon^{\phi\kappa\lambda\tau} \epsilon^{\nu\rho\delta\tau} \left(g_{\kappa\rho} \, g_{\lambda\delta} - g_{\kappa\delta} \, g_{\lambda\rho} \right) = -12 \, g_{\phi\nu}. \tag{B.28}$$

The term $\lambda^a \lambda^b \delta^{ab} = 16$, so that putting all together $\Pi_V(q^2)$ becomes

$$\Pi_V(q^2)|_{\langle G^2\rangle} = -i \frac{g^2}{3q^2} \langle G^2\rangle \int \frac{d^4k}{(2\pi)^4} \frac{k^\mu (k-q)_\mu}{k^4 (k-q)^4}. \tag{B.29}$$

The integral is found in Eqs. (A.23)–(A.24), and after substitution the vector correlator is given by

$$\Pi_V(q^2)|_{\langle G^2\rangle} = \frac{1}{12} \frac{1}{q^4} \langle \frac{\alpha_s}{\pi} G^2 \rangle, \tag{B.30}$$

where $\alpha_s \equiv g^2/(4\pi)$.

There is an alternative procedure to obtain this result using the free quark propagators in the loop and coupling two gluon lines ending in the vacuum in all possible (three) ways. This calculation is far more lengthy than the one using the "dressed" quark propagator, Eq. (B.25), as above.

In the case of the heavy-quark vector current correlator, Eq. (B.1), now with $J_\mu(x) = \bar{Q}(x)\gamma_\mu Q(x)$, where $Q = c, b$, the gluon condensate term in the OPE is obtained as above, except that the quark mass must be retained in the dressed quark propagator. This is given in Eq. (A.1). There are two types of contributions remaining after performing the traces. These are (i) the product of the two terms linear in $G_{\mu\nu}$, and (ii) the product of the quadratic term, G^2, and the quark term. The former involves the integral

$$I_1(q^2) = \int \frac{d^4k}{(2\pi)^4} \frac{k^\mu (k-q)_\mu}{(k^2 - m_Q^2)^2 \, [(k-q)^2 - m_Q^2]^2} \tag{B.31}$$

and the latter involves

$$I_2(q^2) = \int \frac{d^4k}{(2\pi)^4} \frac{k^\mu (k-q)_\mu - 2k^2}{(k^2 - m_Q^2)^4 \, [(k-q)^2 - m_Q^2]}, \tag{B.32}$$

which are given in Eqs. (A.23)–(A.24). The result for the gluon condensate contribution to Π_V in the heavy-quark sector is [1]

$$\Pi_V(q^2)|_{\langle G^2\rangle} = \frac{1}{48} \frac{1}{q^4} \langle \frac{\alpha_s}{\pi} G^2 \rangle \left[\frac{3(1+a)(1-a)^2}{a^2} \right.$$
$$\left. \times \frac{1}{2\sqrt{a}} \ln \left(\frac{\sqrt{a}+1}{\sqrt{a}-1} \right) - \frac{3a^2 - 2a + 3}{a^2} \right]. \tag{B.33}$$

Another important correlator in the heavy-quark sector is the pseudoscalar current correlator, defined by

$$\Pi_5(q^2) = i \int d^4x \, e^{iq\cdot x} \langle 0|T(J_5(x) J_5^\dagger(0))|0\rangle, \tag{B.34}$$

where

$$J_5(x) = \bar{Q}(x) i \gamma_5 Q(x), \tag{B.35}$$

with $Q(x)$ a heavy-quark (charm/bottom) field. The gluon condensate contribution to $\Pi_5(q^2)$ can be obtained following the procedure for the vector current correlator. The result is given by

$$\Pi_5(q^2)|_{\langle G^2 \rangle} = \frac{1}{48} \frac{1}{4 m_Q^2} \left\langle \frac{\alpha_s}{\pi} G^2 \right\rangle \left[\frac{3(1 + 3a)(1 - a)^2}{a^2} \right.$$
$$\left. \times \frac{1}{2\sqrt{a}} \ln \left(\frac{\sqrt{a} + 1}{\sqrt{a} - 1} \right) - \frac{9 a^2 + 4 a + 3}{a^2} \right]. \tag{B.36}$$

Finally, the gluon condensate contribution to the pseudoscalar current correlator, $\psi_5(q^2)$, Eq. (1.5) is obtained along the same lines as for the vector current correlator. The result is

$$\psi_5(q^2)|_{\langle G^2 \rangle} = -(m_u + m_d)^2 \frac{1}{8 q^2} \left\langle \frac{\alpha_s}{\pi} G^2 \right\rangle. \tag{B.37}$$

Regarding radiative corrections to current correlators, the first order in α_s is the only one doable *by hand*, and even then its derivation involves a couple of pages of intermediate steps. Hence, only the general outline is presented here. Considering the electrically charged, light quark, vector current correlator, Eq. (B.1), at the one-loop level with one-gluon exchange

$$\Pi_{\mu\nu}^V(q^2) = i^3 \int d^4x \, d^4y \, d^4z \, e^{iq \cdot x} \langle 0|T \left(V_\mu(x) \, V_\nu^\dagger(0) L_I(y) \, L_I(z) \right) |0 \rangle, \tag{B.38}$$

where the quark-gluon interaction Lagrangian is

$$L_I(y) = g \, \bar{u}^\alpha(y)|_m \, \gamma^\lambda|_{mr} \left(\frac{\lambda^a}{2} \right)^{\alpha\beta} u^\beta(y)|_r \, G_\lambda^a(y), \tag{B.39}$$

and similarly for $L_I(z)$. Substituting the Lagrangians into Eq. (B.38) gives rise to three terms corresponding to the three possible ways of connecting quarks with a gluon in the loop, i.e. an up-quark line dressed with a gluon, a down-quark line dressed with a gluon, and an up- and a down-quark lines exchanging a gluon. In the chiral limit the first two give the same answer, so only two terms remain. After integration over the variable x, there remain two four-momentum integrals leading to the simple result

$$\Pi_{\mu\nu}^V(q^2) = (-g_{\mu\nu} q^2 + q_\mu q_\nu) \frac{1}{4 \pi^2} \ln(-q^2/\mu^2) \left(1 + \frac{\alpha_s}{\pi} \right). \tag{B.40}$$

Radiative corrections to the heavy-quark vector current correlator are important in applications, e.g. in the determination of the heavy-quark masses of the charm and the bottom quark. These have already been discussed in Chap. 9.

Appendix C
Light-Quark Pseudoscalar Current Correlator in QCD

The pseudoscalar current correlator in the up-down-quark sector $\psi_5(q^2)$, Eq. (1.5), is given by

$$\psi_5(q^2) = (\bar{m}_u + \bar{m}_d)^2 \left\{ - q^2 \, \Pi_0(q^2) + (\bar{m}_u + \bar{m}_d)^2 \, \Pi_2(q^2) \right.$$

$$\left. - \frac{C_q}{-q^2} (\bar{m}_u + \bar{m}_d) \langle \bar{q} \, q \rangle + \frac{C_4 \langle O_4 \rangle}{-q^2} + \mathcal{O}\left(\frac{1}{q^4}\right) \right\}, \qquad (C.1)$$

where $\bar{m}_q$ stands for the quark mass in the MS-bar renormalization scheme. In the strange-quark case one must perform the replacement $(\bar{m}_u + \bar{m}_d) \Rightarrow (\bar{m}_q + \bar{m}_s)$, with $\bar{m}_q \equiv (\bar{m}_u + \bar{m}_d)/2$. The coefficient C_q above is $C_q = 1/2$. Of particular importance is to notice the asymptotic behaviour of $\psi_5(q^2)$, as it diverges quadratically

$$\lim |_{-q^2 \to \infty} \psi_5(q^2) \sim q^2. \qquad (C.2)$$

This implies that dispersion relations for $\psi_5(q^2)$ are to involve two subtractions. In the framework of FESR this is usually of no major concern as integrals are over a finite range. Otherwise, the alternative is to consider the second derivative of $\psi_5(q^2)$, to be discussed at the end of this Appendix.

The perturbative QCD function $\Pi_0(q^2)$ is known up to order $\mathcal{O}(\alpha_s^4)$ [7–10], to wit.

$$\Pi_0(q^2) = \frac{1}{16\,\pi^2} \left[- 12 + 6L + a_s A_1(q^2) + a_s^2 A_2(q^2) + a_s^3 A_3(q^2) + a_s^4 A_4(q^2) \right], \qquad (C.3)$$

where $L \equiv \ln(-q^2/\mu^2)$, $a_s \equiv \alpha_s(-q^2)/\pi$, and the $A_i(q^2)$ terms are given by

$$A_1(q^2) = -\frac{131}{2} + 34\,L - 6\,L^2 + 24\,\zeta(3), \tag{C.4}$$

$$A_2(q^2) = \left(4\,n_F\,\zeta(3) - \frac{65}{4}n_F - 117\,\zeta(3) + \frac{10801}{24}\right)L + \left(\frac{11}{3}n_F - 106\right)L^2$$
$$+ \left(-\frac{n_F}{3} + \frac{19}{2}\right)L^3 + \text{constants}, \tag{C.5}$$

$$A_3(q^2) = C_1\,L - 6\left(\frac{4781}{18} - \frac{475}{8}\,\zeta(3)\right)L^2 + 229\,L^3 - \frac{221}{16}\,L^4, \tag{C.6}$$

where

$$C_1 = \frac{4748953}{864} - \frac{\pi^4}{6} - \frac{91519}{36}\,\zeta(3) + \frac{715}{2}\,\zeta(5), \tag{C.7}$$

and finally

$$A_4(q^2) = \sum_{i=1}^{5} H_i\,L^i, \tag{C.8}$$

where the coefficients H_i contain a very large number of terms, e.g. H_1 has twenty-three terms!. Hence, we only list their numerical values [10]: $H_1 = 33532.3$, $H_2 = -15230.645111$, $H_3 = 3962.454926$, $H_4 = -534.0520833$, and $H_5 = 24.17187500$.

The remaining terms in Eq. (C.1) are

$$\Pi_2(q^2) = \frac{1}{16\,\pi^2}\left\{-12 + 12\,L + a_s\left[-100 + 64\,L - 24\,L^2 + 48\,\zeta(3)\right]\right\}, \tag{C.9}$$

$$C_q = \frac{1}{2} + \left[\frac{7}{3} - L\right]a_s, \tag{C.10}$$

$$C_4\langle O_4\rangle = -\frac{1}{8}\,a_s\,\langle G_{\mu\nu}, G_{\mu\nu}\rangle\,(1 + \mathcal{O}(a_s)), \tag{C.11}$$

where the radiative correction, $\mathcal{O}(a_s)$, is not known. The current correlator in the strange-quark case follows from the above results after replacing $\bar{m}_d$ by $\bar{m}_s$, and $\bar{m}_u$ by $(\bar{m}_u + \bar{m}_d)/2$.

Returning to the issue of the asymptotic behaviour of $\psi_5(q^2)$, in FOPT there is no difference between using in FESR the function $\psi_5(q^2)$ or its second derivative, $\psi_5''(q^2)$. This is due to the identity, Eq. (4.3), plus the fact that the hadronic term of the FESR involving $\psi_5(q^2)$ is identical to the one involving its second derivative, $\psi_5''(q^2)$. This is not necessarily the case in CIPT. Hence, one might need the PQCD expression of $\psi_5''(q^2)$. Up to order $\mathcal{O}(\alpha_s^4)$, after Renormalization Group Improvement (RGI), i.e. setting the renormalization scale $\mu^2 = -q^2$, it is as follows [7–10]

$$\psi_5''(q^2)|_{PQCD} = -\frac{(m_u + m_d)^2}{16\,\pi^2}\frac{1}{s}\sum_{m=0} K_m\,a_s^m, \tag{C.12}$$

where the coefficients K_m are

$$K_0 = 6, \tag{C.13}$$

$$K_1 = 22, \tag{C.14}$$

$$K_2 = 4n_f \zeta(3) - \frac{65}{4} n_f - 117\zeta(3) + \frac{10801}{24} + 2\left(\frac{11}{3} n_f - 106\right), \tag{C.15}$$

$$\begin{aligned}
K_3 = & \frac{6163613}{864} - \frac{109735}{36}\zeta(3) + \frac{815}{2}\zeta(5) + \left[-\frac{46147}{81} + \frac{524}{3}\zeta(3)\right. \\
& \left. - 5\zeta(4) - \frac{50}{3}\zeta(5)\right]n_f + \left[\frac{15511}{1944} - 2\zeta(3)\right]n_f^2 - \frac{49349}{12} + \frac{1755}{2}\zeta(3) \\
& + \left[\frac{11651}{36} - 59\zeta(3)\right]n_f + \left[-\frac{275}{54} + \frac{1}{3}\zeta(3)\right]n_f^2.
\end{aligned} \tag{C.16}$$

$$K_4 = C_{41} + 2C_{42}, \tag{C.17}$$

where

$$\begin{aligned}
C_{41} = & \frac{10811054729}{82944} - \frac{3887351}{54}\zeta(3) + \frac{458425}{72}\zeta^2(3) + \frac{265}{3}\zeta(4) \\
& + \frac{373975}{72}\zeta(5) - \frac{4125}{16}\zeta(6) - \frac{178045}{128}\zeta(7) + \left[-\frac{1045811915}{62208}\right. \\
& + \frac{5747185}{864}\zeta(3) - \frac{2865}{8}\zeta^2(3) - \frac{9131}{96}\zeta(4) + \frac{41215}{72}\zeta(5) \\
& + \frac{2875}{48}\zeta(6) + \frac{665}{12}\zeta(7)\Big]n_f + \left[\frac{220313525}{373248} - \frac{11875}{72}\zeta(3) + 5\zeta^2(3)\right. \\
& + \frac{25}{16}\zeta(4) - \frac{5015}{72}\zeta(5)\Big]n_f^2 + \left[-\frac{520771}{93312} + \frac{65}{72}\zeta(3) + \frac{1}{24}\zeta(4)\right. \\
& + \frac{5}{3}\zeta(5)\Big]n_f^3 = 56824.55903 - 8725.6816\,n_f + 328.69544\,n_f^2 \\
& - 2.722464877\,n_f^3,
\end{aligned} \tag{C.18}$$

which for $n_f = 3$ gives: $C_{41} = 33532.27$. The second term in Eq. (C.17) is given by

$$\begin{aligned}
C_{42} = & -\frac{49573615}{1152} + \frac{535759}{32}\zeta(3) - \frac{30115}{16}\zeta(5) + \left[\frac{56935973}{10368}\right. \\
& - \frac{243511}{144}\zeta(3) + 5\zeta(4) + \frac{1115}{8}\zeta(5)\Big]n_f + \left[-\frac{6209245}{31104} + \frac{500}{9}\zeta(3)\right. \\
& - \frac{25}{6}\zeta(5)\Big]n_f^2 + \left[\frac{985}{486} - \frac{5}{9}\zeta(3)\right]n_f^3,
\end{aligned} \tag{C.19}$$

which for $n_f = 3$ gives $C_{42} = -15230.646$.

Turning to the non-perturbative expression of $\psi_5''(q^2)$, the gluon condensate contribution is given by

$$\psi_5''(q^2)|_{\langle G^2\rangle} = -\frac{1}{4}\,(\bar{m}_u + \bar{m}_d)^2\,\frac{1}{(q^2)^3}\,\left\langle\frac{\alpha_s}{\pi}\,G^2\right\rangle. \qquad (C.20)$$

In principle, there is an issue with the gluon condensate contribution to light-quark correlators, related to the removal of logarithmic quark-mass singularities [11]. However, this turns out to be numerically unimportant, given the size of uncertainties from other sources, so that the above expression can be safely used. The next power correction is that involving the quark condensate, given by

$$\psi_5''(q^2)|_{\langle \bar{q}\,q\rangle} = 2\,\frac{(\bar{m}_u + \bar{m}_d)^2}{(q^2)^3}\,(\bar{m}_u + \bar{m}_d)\langle \bar{q}\,q\rangle\left(\frac{1}{2} + \frac{7}{3}\,\frac{\alpha_s}{\pi}\right). \qquad (C.21)$$

The basic CIPT integrals are formally given in Eqs. (4.7)–(4.8), and will be discussed in the next Appendix.

Appendix D
QCD Integrals of Light-Quark Pseudoscalar Current Correlator in QCD

Starting with integration in FOPT, the contour integral in the FESR for the QCD light-quark current correlator involving an (analytic) integration kernel of the form

$$\Delta_5(s) = 1 - a_0\, s - a_1\, s^2, \tag{D.1}$$

is given by

$$\delta_5(s_0)|_{QCD} \equiv -\frac{1}{2\pi i} \oint_{C(|s_0|)} ds\, \Delta_5(s)\, \psi_5(s)|_{QCD}, \tag{D.2}$$

$$\delta_5(s_0)|_{1-LOOP} = \frac{\overline{m}^2(s_0)}{16\pi^2}\, C_{01} \left[\frac{s_0^2}{2} - a_0\, \frac{s_0^3}{3} - a_1\, \frac{s_0^4}{4} \right], \tag{D.3}$$

$$\delta_5(s_0)|_{2-LOOP} = \frac{\overline{m}^2(s_0)}{16\pi^2}\, \frac{\alpha_s(s_0)}{\pi} \left[C_{11} \left(\frac{s_0^2}{2} - a_0\, \frac{s_0^3}{3} - a_1\, \frac{s_0^4}{4} \right) \right.$$
$$\left. - 2\, C_{12} \left(\frac{s_0^2}{4} - a_0\, \frac{s_0^3}{9} - a_1\, \frac{s_0^4}{16} \right) \right], \tag{D.4}$$

$$\delta_5(s_0)|_{3-LOOP} = \frac{\overline{m}^2(s_0)}{16\pi^2} \left[\frac{\alpha_s(s_0)}{\pi} \right]^2 \left\{ C_{21} \left(\frac{s_0^2}{2} - a_0\, \frac{s_0^3}{3} - a_1\, \frac{s_0^4}{4} \right) \right.$$
$$- 2\, C_{22} \left(\frac{s_0^2}{4} - a_0\, \frac{s_0^3}{9} - a_1\, \frac{s_0^4}{16} \right) - 6\, C_{23} \left[\frac{s_0^2}{2} \left(\frac{\pi^2}{6} - \frac{1}{4} \right) \right.$$
$$\left. \left. - a_0\, \frac{s_0^3}{3} \left(\frac{\pi^2}{6} - \frac{1}{9} \right) - a_1\, \frac{s_0^4}{4} \left(\frac{\pi^2}{6} - \frac{1}{16} \right) \right] \right\}, \tag{D.5}$$

C. A. Dominguez, *Quantum Chromodynamics Sum Rules*, SpringerBriefs in Physics, https://doi.org/10.1007/978-3-319-97722-5

$$\delta_5(s_0)|_{4-LOOP} = \frac{\overline{m}^2(s_0)}{16\pi^2} \left[\frac{\alpha_s(s_0)}{\pi}\right]^3 \left\{ C_{31}\left(\frac{s_0^2}{2} - a_0 \frac{s_0^3}{3} - a_1 \frac{s_0^4}{4}\right) \right.$$

$$- 2\, C_{32}\left(\frac{s_0^2}{4} - a_0 \frac{s_0^3}{9} - a_1 \frac{s_0^4}{16}\right) - 6\, C_{33}\left[\frac{s_0^2}{2}\left(\frac{\pi^2}{6} - \frac{1}{4}\right)\right.$$

$$\left. - a_0 \frac{s_0^3}{3}\left(\frac{\pi^2}{6} - \frac{1}{9}\right) - a_1 \frac{s_0^4}{4}\left(\frac{\pi^2}{6} - \frac{1}{16}\right)\right] + 24\, C_{34}\left[\frac{s_0^2}{2}\right.$$

$$\left.\left. \times \left(\frac{\pi^2}{6} - \frac{1}{4}\right) - a_0 \frac{s_0^3}{9}\left(\frac{\pi^2}{6} - \frac{1}{9}\right) - a_1 \frac{s_0^4}{16}\left(\frac{\pi^2}{6} - \frac{1}{16}\right)\right]\right\}. \quad (D.6)$$

$$\delta_5(s_0)|_{5-LOOP} = \frac{\overline{m}^2(s_0)}{16\pi^2} \left[\frac{\alpha_s(s_0)}{\pi}\right]^4 \left\{ C_{41}\left(\frac{s_0^2}{2} - a_0 \frac{s_0^3}{3} - a_1 \frac{s_0^4}{4}\right) \right.$$

$$- 2\, C_{42}\left(\frac{s_0^2}{4} - a_0 \frac{s_0^3}{9} - a_1 \frac{s_0^4}{16}\right) - 6\, C_{43}\left[\frac{s_0^2}{2}\left(\frac{\pi^2}{6} - \frac{1}{4}\right)\right.$$

$$\left. - a_0 \frac{s_0^3}{3}\left(\frac{\pi^2}{6} - \frac{1}{9}\right) - a_1 \frac{s_0^4}{4}\left(\frac{\pi^2}{6} - \frac{1}{16}\right)\right] + 24\, C_{44}\left[\frac{s_0^2}{4}\right.$$

$$\left. \times \left(\frac{\pi^2}{6} - \frac{1}{4}\right) - a_0 \frac{s_0^3}{9}\left(\frac{\pi^2}{6} - \frac{1}{9}\right) - a_1 \frac{s_0^4}{16}\left(\frac{\pi^2}{6} - \frac{1}{16}\right)\right]$$

$$+ 120\, C_{45}\left[\frac{s_0^2}{2}\left(\frac{\pi^4}{120} - \frac{\pi^2}{24} + \frac{1}{16}\right) - a_0 \frac{s_0^3}{3}\left(\frac{\pi^4}{120}\right.\right.$$

$$\left.\left.\left. - \frac{\pi^2}{54} + \frac{1}{81}\right) - a_1 \frac{s_0^4}{4}\left(\frac{\pi^4}{120} - \frac{\pi^2}{96} - \frac{1}{256}\right)\right]\right\}, \quad (D.7)$$

where $\overline{m} \equiv \overline{m}_u + \overline{m}_d$, and the constants C_{ij} above, for three quark flavours, are:
$C_{01} = 6$, $\quad C_{11} = 34$, $\quad C_{12} = -6$, $\quad C_{21} = -105\, \zeta(3) + 9631/24$, $\quad C_{22} = -95$,
$C_{23} = 17/2$, $C_{31} = 4748953/864 - \pi^4/6 - 91519\, \zeta(3)/36 + 715\, \zeta(5)/2$, $C_{32} = -6\,[4781/18 - 475\, \zeta(3)/8]$, $C_{33} = 229$, $C_{34} = -221/16$, $C_{41} = 33532.26$, $C_{42} = -15230.6451$, $C_{43} = 3962.45493$, $C_{44} = -534.052083$, $C_{45} = 24.1718750$. The leading non-perturbative contributions are due to the gluon and the light-quark condensates, which give

$$\delta_5(s_0)|_{<G^2>} = \frac{\overline{m}^2(s_0)}{8} \langle\frac{\alpha_s}{\pi}G^2\rangle \left[1 + \frac{\alpha_s(s_0)}{\pi}\left(\frac{11}{2} + 2\, a_0\, s_0 + a_1\, s_0^2\right)\right], \quad (D.8)$$

$$\delta_5(s_0)|_{<\bar{q}q>} = \overline{m}^2(s_0)\, \langle m_{ud}\, \bar{q}\, q\rangle \left[1 + \frac{\alpha_s(s_0)}{\pi}\left(\frac{14}{3} + 2\, a_0\, s_0 + a_1\, s_0^2\right)\right], \quad (D.9)$$

where $\langle\bar{u}u\rangle = \langle\bar{d}d\rangle \equiv \langle\bar{q}q\rangle$ is a very good approximation. The case of the strange-quark mass is obtained as above after redefining $\overline{m} \rightarrow (\overline{m}/2 + \overline{m}_s)$, .

Turning to CIPT, integrating Eq. (C.12) gives

$$\delta_5(s_0)|^{RGI}_{PQCD} = \frac{\bar{m}^2}{16\pi^2} \sum_{n=0}^{4} K_n \frac{1}{2\pi i} \oint_{C(|s_0|)} \frac{ds}{s} \left[F(s) - F(s_0) \right] \left(\frac{\bar{\alpha}_s(s)}{\pi} \right)^n, \quad \text{(D.10)}$$

where

$$F(s) = \sum_{N=1}^{4} b_N s^N, \quad \text{(D.11)}$$

with

$$b_1 = -\left[s_0 - a_0 \frac{s_0^2}{2} - a_1 \frac{s_0^3}{3} \right], \quad b_2 = \frac{1}{2}, \quad b_3 = -\frac{a_0}{6}, \quad b_4 = -\frac{a_1}{12}. \quad \text{(D.12)}$$

Finally, Eq. (D.10)becomes

$$\delta_5(s_0)|^{RGI}_{PQCD} = \frac{\bar{m}^2}{16\pi^2} \sum_{M} K_M \left[\sum_{N} b_N \frac{1}{2\pi i} \oint_{|s_0|} \frac{ds}{s} s^N \left(\frac{\bar{\alpha}_s(s)}{\pi} \right)^M - F(s_0) \frac{1}{2\pi i} \right.$$

$$\left. \times \oint_{|s_0|} \frac{ds}{s} \left(\frac{\bar{\alpha}_s(s)}{\pi} \right)^M \right] = \frac{\bar{m}^2}{16\pi^2} \sum_{M} K_M \left[\sum_{N} b_N I^a_{NM} - F(s_0) I^b_{NM} \right], \quad \text{(D.13)}$$

where the two basic integrals, $I^a_{NM} \equiv I^a_{NM}(s_0)$ and $I^b_{NM} \equiv I^b_{NM}(s_0)$, were introduced in Eqs. (4.7)–(4.8).

Appendix E
QCD Thermal Space-like Spectral Functions

This Appendix deals with the QCD expression of the so-called *scattering term* for a vector current correlation function of non-zero, equal mass quarks. It is actually a space-like contribution. Other current correlators may be treated similarly. The vector current is $V_\mu(x) =: \overline{Q}^a(x)\, \gamma_\mu\, Q^a(x) :$, where $Q(x)$ is a quark field of mass m_Q, and a is the colour index. In the time-like region one has

$$\Pi^a_{\mu\nu}(q^2) \equiv (-g_{\mu\nu}\, q^2 + q_\mu q_\nu)\, \Pi^a(q^2) = -i^3\, N_c \int d^4x\; e^{iqx}\, Tr\left[\gamma_\mu\, S_F(x)\, \gamma_\nu\, S_F(-x)\right], \quad (E.1)$$

where $S_F(x)$ is the quark propagator in space-time, and $N_c = 3$. Transforming the propagators to momentum-space, performing the integrations, and taking the imaginary part of $\Pi(q^2)$ gives

$$Im\; \Pi^a(q^2) = \frac{3}{16\,\pi} \int_{-v}^{+v} dx\; (1 - x^2) = \frac{1}{8\,\pi}\, v(3 - v^2), \quad (E.2)$$

where the variable $v \equiv v(q^2)$ is given by

$$v(q^2) = \left(1 - \frac{4\, m_Q^2}{q^2}\right)^{1/2}. \quad (E.3)$$

Notice that the normalization factor of $\Pi^a(q^2)$ for massless quarks is $Im\; \Pi^a(q^2) = 1/(4\,\pi)$ (Fig. E.1).

The extension to finite T can be performed using the Dolan-Jackiw thermal propagator, Eq. (12.3), in Eq. (E.1), to obtain.

$$Im\; \Pi^a(q^2, T) = \frac{3}{16\,\pi} \int_{-v}^{+v} dx\; (1 - x^2) \left[1 - n_F\left(\frac{|\mathbf{q}|x + \omega}{2T}\right) - n_F\left(\frac{|\mathbf{q}|x - \omega}{2T}\right)\right]. \quad (E.4)$$

© The Author(s), under exclusive licence to Springer Nature Switzerland AG 2018
C. A. Dominguez, *Quantum Chromodynamics Sum Rules*,
SpringerBriefs in Physics, https://doi.org/10.1007/978-3-319-97722-5

Fig. E.1 The complex energy, ω-plane, showing the central cut around the origin (scattering term), extending between $\omega = -|\mathbf{q}|$, and $\omega = |\mathbf{q}|$. The standard (time-like) annihilation right-hand and left-hand cuts at $\omega = \pm\left[|\mathbf{q}|^2 + \omega_{th}^2\right]^{1/2}$ are not shown (ω_{th} is some channel dependent mass threshold)

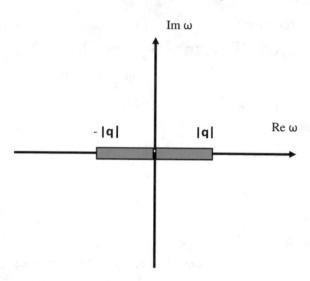

In the rest frame of the thermal medium, $|\mathbf{q}| \rightarrow \mathbf{0}$, this reduces to

$$Im\ \Pi^a(\omega, T) = \frac{3}{16\pi} \int_{-v}^{+v} dx\ (1 - x^2)\ [1 - 2\,n_F\,(\omega/2T)]$$

$$= \frac{3}{16\pi} \int_{-v}^{+v} dx\ (1 - x^2)\tanh\left(\frac{\omega}{4T}\right). \tag{E.5}$$

Proceeding to the scattering term, the equivalent to Eq. (E.4) is

$$Im\ \Pi^s(q^2, T) = \frac{3}{8\pi} \int_{v}^{\infty} dx\ (1 - x^2)\left[n_F\left(\frac{|\mathbf{q}|x + \omega}{2T}\right) - n_F\left(\frac{|\mathbf{q}|x - \omega}{2T}\right)\right], \tag{E.6}$$

where the integration limits arise from the bounds in the angular integration in momentum space. Notice that this term vanishes identically at $T = 0$, and the overall multiplicative factor is twice the one in Eq. (E.4). Next, the term in brackets in Eq. (E.6) becomes a derivative

$$Im\ \Pi^s(q^2, T) = \frac{3}{8\pi} \frac{\omega}{T} \int_{v}^{\infty} dx\ (1 - x^2) \frac{d}{dy} n_F(y) \tag{E.7}$$

where $y = |\mathbf{q}|\, x/(2\,T)$. This expression reduces to

$$Im\ \Pi^s(q^2, T) = \frac{3}{4\pi} \frac{\omega}{|\mathbf{q}|}\left[-n_F\left(\frac{|\mathbf{q}|\,v}{2T}\right)(1 - v^2) + \frac{8\,T^2}{|\mathbf{q}|^2} \int_{|\mathbf{q}|\,v/2T}^{\infty} y\,n_F(y)\,dy\right]. \tag{E.8}$$

In the limit $|\mathbf{q}| \to 0$ this result becomes

$$Im\ \Pi^s(q^2, T) = \frac{3}{\pi} \lim_{\substack{|\mathbf{q}|\to 0 \\ \omega \to 0}} \frac{\omega}{|\mathbf{q}|^3}\ m_Q^2 \left[n_F\left(\frac{m_Q}{T}\right) + \frac{2\,T^2}{m_Q^2} \int_{m_Q/T}^{\infty} y\,n_F(y)\,dy \right].$$
(E.9)

After careful performance of the limit, in the order indicated, the singular term $\omega/|\mathbf{q}|^3$ above becomes a delta function

$$\lim_{\substack{|\mathbf{q}|\to 0 \\ \omega \to 0}} \frac{\omega}{|\mathbf{q}|^3} = \frac{2}{3}\,\delta(\omega^2),$$
(E.10)

and the final result for the scattering term is

$$Im\ \Pi^s(\omega, T) = \frac{2}{\pi}\,m_Q^2\,\delta(\omega^2) \left[n_F\left(\frac{m_Q}{T}\right) + \frac{2\,T^2}{m_Q^2} \int_{m_Q/T}^{\infty} y\,n_F(y)\,dy \right].$$
(E.11)

Depending on the correlator, the limiting function, Eq. (E.10), could instead be less singular in $|\mathbf{q}|$, in which case the scattering term would vanish identically.

References

1. V.A. Novikov, M.A. Shifman, A.I. Vainshtein, V.I. Zakharov, Fortschritte der Physyk **32**, 585 (1984)
2. M.A. Shifman, A.I. Vainshtein, V.I. Zakharov, Nucl. Phys. B **147**, 385 (1979), *ibid.*, B **147**, 448 (1979)
3. A. Gomez Nicola, J.R. Pelaez, J. Ruiz de Elvira, Phys. Rev. D **82**, 074012 (2010)
4. S. Bodenstein, C.A. Dominguez, S.I. Eidelman, H. Spiesberger, K. Schilcher, J. High Energy Phys. **01**, 039 (2012)
5. C.A. Dominguez, L.A. Hernandez, K. Schilcher, H. Spiesberger, J. High Energy Phys. **03**, 053 (2015)
6. H. Pietschmann, *Weak Interactions-Formulae, results, and Derivations*, Springer Verlag/Wien (1983)
7. S.G. Gorishnii, A.L. Kataev, S.A. Larin, Phys. Lett. B **259**, 144 (1991)
8. K. Chetyrkin, B.A. Kniehl, M. Steinhauser, Phys. Rev. Lett. **79**, 2184 (1997)
9. P.A. Baikov, K.J. Chetyrkin, J.H. Kühn, Phys. Rev. Lett. **96**, 012003 (20016)
10. K.J. Chetyrkin, Private communication
11. K.G. Chetyrkin, C.A. Dominguez, D. Pirjol, K. Schilcher, Phys. Rev. D 51, 5090 (1995); K.G. Chetyrkin, D. Pirjol, K. Schilcher, Phys. Lett. B 404, 337 (1997). C.A. Dominguez, L. Pirovano, K. Schilcher. Phys. Lett. B **425**, 193 (1998)

Printed in the United States
By Bookmasters